T0261173

SPIRIT WHALES AND SLOTH TALES

Published in association with the Burke Museum

Spirit
Whales

FOSSILS OF WASHINGTON STATE

Elizabeth A. Nesbitt and David B. Williams

Sloth
& Tales

University of Washington Press   *Seattle*

To Mick, who always stood by me. LIZ

To Marjorie, You rock my world! DAVID

A RUTH KIRK BOOK

Spirit Whales and Sloth Tales was published with the assistance
of a grant from the Ruth Kirk Book Fund, which supports publications
that inform the general public on the history, natural history, archaeology,
and Native cultures of the Pacific Northwest.

Copyright © 2023 by the University of Washington Press
Design by Mindy Basinger Hill
Composed in Calluna Pro and Calluna Sans Pro
All photos by Michael Rich unless otherwise stated.

27   26   25   24   23   5   4   3   2   1

Printed and bound in the United States of America

All rights reserved. No part of this publication may be reproduced or
transmitted in any form or by any means, electronic or mechanical,
including photocopy, recording, or any information storage or retrieval
system, without permission in writing from the publisher.

UNIVERSITY OF WASHINGTON PRESS   *uwapress.uw.edu*
Published in association with the Burke Museum *www.burkemuseum.org*

Cataloging information is available from the Library of Congress
ISBN 9780295752327 (paperback)
ISBN 9780295752334 (ebook)

∞ This paper meets the requirements of ANSI/NISO Z39.48-1992
(Permanence of Paper).

CONTENTS

PREFACE

Genius and science have burst the limits of space and few observations,
explained by just reasoning, have unveiled the mechanisms of the universe.
Would it not also be glorious for man to burst the limits of time and
by a few observations, to ascertain the history of this world, and the series
of events which preceded the birth of the human race?
BARON GEORGES LEOPOLD CHRÊTIEN FRÉDÉRIC DAGOBERT CUVIER
Essay on the Theory of the Earth

In 2014, a construction worker near downtown Seattle found a tusk while excavating for a new building. Paleontologists from the Burke Museum of Natural History and Culture soon arrived. They worked past midnight to unearth the delicate fossil, wrap it in protective plaster, and arrange transport to the museum for what turned out to be the remains of a mammoth that lived around 15,000 years ago. Also arriving were the media and onlookers, some of whom chanted, "Dig it up! Dig it up!"

For those who watched and participated in the discovery at South Lake Union, it was a thrilling time as the fossil remains connected viewers to a distant world recently scraped raw by a 3,000-foot-thick sheet of ice. Seeing the tusk, they could imagine a chilly postglacial scene of giant mammals, formidable predators, and plants recolonizing a barren landscape. Like all fossils, the tusk was a window into a past completely different from the urban one now found at this location. The onlookers who experienced those heady few hours are certainly not alone in succumbing to the allure of ancient time; over the decades uncounted numbers of people have had similar experiences, finding a fossil in Washington State and being tantalized by it and the connections it provides to the region's geologic history.

No matter where you wander in Washington, you are never very far from the past and the fossil evidence of those that came before. You can find trilobites near the Idaho border, primitive horses on the Columbia Plateau, exquisite flowers in Republic, giant bird tracks near Bellingham, and curious bearlike beasts on the Olympic Peninsula.

With abundant and well-exposed rock layers, Washington has fossils dating from Ice Age mammals only 12,000 years old back to marine invertebrates more than 500 million years old.

The goal of *Spirit Whales and Sloth Tales* is to share the paleontological stories, to provide insights into ancient natural history, and to help people understand the dynamic ecosystems of extinct plants and animals. This book also aims to highlight those who found the fossils. Many were discovered by paleontologists but, as happened with the Seattle mammoth, numerous fossils have been found by nonprofessionals, people who were simply observant and paying attention to the natural world around them. Including their names in *Spirit Whales and Sloth Tales* acknowledges and honors their contributions to the story of paleontology in the state, for without them, the stories of our past would be far less interesting.

My more than two decades of work at the Burke Museum in Seattle exemplifies these contributions. I have had the good fortune to work with an astounding crew of volunteer fossil enthusiasts. Four of them—Ross Berglund, Wes Wehr, and Jim and Gail Goedert—have won the prestigious Harrell L. Strimple Award, given annually by the national Paleontological Society for outstanding achievement in paleontology by an amateur. What unites this quartet is their passion for fossils and devotion to sharing what they find with the public. You'll meet them later in the book.

I was also fortunate at the Burke to work with and get to know writer and naturalist David B. Williams, who is my coauthor. He brings a strong background in geology and paleontology, including extensive writing about the Pacific Northwest in which he focuses on connecting people to the natural world around them. David's books include the award-winning *Homewaters: A Human and Natural History of Puget Sound* and *Stories in Stone: Travels through Urban Geology*. I further want to acknowledge Mike Rich, a Burke volunteer whose beautiful and detailed fossil photographs are essential to this book.

Both of us have long been inspired by visitors to the Burke and their passion for our state's fossils, and it was always a pleasure to see their joy when we could identify a fossil or provide details about the environment where the plant or animal lived. In telling the more exciting stories of the state's fossils in this book, we hope that we convey the excitement we saw in museum visitors, as well as the excitement we, and many others, have experienced in finding fossils in the field.

Working with the public at the Burke and teaching undergraduate classes at the University of Washington and other colleges, I have come to appreciate the stories and the fossils that interest nonscientists, which has also has helped me choose the fossils to include in this book. They are the fossils that most intrigued and excited the Burke's visitors, be they amateur fossil hounds, children with a budding interest in paleontology, or adults inspired by the natural world, which are our primary audiences. We have not attempted to be complete but have included the significant fossils and ones that we believe will further spark your curiosity to get out and explore yourself.

In order to help readers more fully appreciate what fossils have to tell us, we start the book with a basic introduction to geology, and specifically the geology of Washington State, along with an understanding of how plants and animals fossilize and how geologists determine the age of fossils. We follow with twenty-four profiles that are the heart and most important part of the book. We see each of these profiles as a short story, where we can explore a topic that may or may not relate to the preceding or following profile but that fits into the broader story of life over time in Washington.

Like a paleontologist excavating fossils, the profiles are organized with the youngest first, allowing the reader to dig deeper and deeper, unearthing stories, strata by strata. Each profile focuses on a specific plant, animal, or environment, often weaving in human history and geology, and always with a goal of fleshing out details necessary for a deeper understanding that will help make the fossils come to life. Ultimately, our goal is for you to come away with a more thorough appreciation of the spectacular paleontology and geology of Washington State.

Many of the fossils mentioned in this book are on display or stored in the Burke Museum's collections; some stories have even been discovered by reexamining fossils long stored in museums. They are also the fossils that I know best illustrate Washington State's complex geology and its changing environment over deep time, from the life-filled warm tropical sea that once covered southwestern Washington to the molten basalt that oozed across Walla Walla to a chilly ocean populated by a surprising diversity of whales unlike any plying today's world. Plus, many simply have wonderful stories and are exquisitely beautiful or uniquely curious.

I have been interested in fossils for most of my life and have been lucky to travel the western United States in search of them, to find them, to dig them up, and to discover their stories. I am still thrilled every time I find one; I know no better way to be inspired by and connect with the natural world around me. My hope is that the stories in this book can help you feel the wonder, passion, and concern I have for fossils of all sizes, from microscopic forms up to the biggest dinosaurs; all have something to teach us and to exhilarate us.

As a working paleontologist, I have seen significant changes in the field, particularly in becoming more interdisciplinary. The careful study of a fossil now includes not only biology and geology but also chemistry, physics, and math. The diverse array of scientists often are taking advantage of new technologies, such as DNA analysis, geochemistry, data modeling, X-ray computerized tomography (CT scans), and 3-D scanning and printing. Not only will you read about these technologies, you will also see that although paleontology is the study of the past, paleontologists have their eyes on the present and how they can help inform the issues of the future, including extinction and climate change. It is truly an exciting time to be a paleontologist and to share with the public the stories that fossils tell, particularly the stories of our state.

With more than a half billion years of history, Washington State has an enviable diversity of fossils. Each is unique. Each interesting. Each tells a story of natural and human history. You don't have to travel to exotic locations to find exciting fossils and do exciting science; it is all right here.

ELIZABETH A. NESBITT

Introduction
Paleontology & Geology

THRILLING TO FIND, EXCITING TO IDENTIFY, and filled with clues of past life on Earth, fossils open up a world of stories about timely and compelling scientific issues such as evolution, climate change, and plate tectonics. In this book, we focus on the fossils of Washington State, journeying close to the present, just 12,000 years ago, when great mammals roamed the land, and diving deep into the past more than 500 million years ago.

Fortunately, many of the fossils described here are in the collections of the Burke Museum of Natural History and Culture on the University of Washington campus in Seattle. Designated in 1899 as the Washington State Museum, it is the "depository for the preservation and exhibition of documents and objects possessing an historical value, of materials illustrating the fauna, flora, anthropology, mineral wealth, and natural resources of the state, and for all documents and objects whose preservation will be of value to the student of history and the natural sciences." As the official repository for fossils for the state of Washington, the Burke employs scientists that have a mandate to disseminate information about the history of life in the state, including paleontologists that focus on all groups of fossil organisms. At present the Burke holds more than 18 million objects that include biological specimens, cultural objects, and fossils.

But what is a fossil? How are organisms preserved in the rocks? How do scientists classify them? And how do we know what we know about them? In order to understand these questions, it helps to understand what paleontology and geology are and how they are related.

The study of evidence of life preserved in rocks is called paleontology, and every paleontologist needs a basic understanding of geology, which provides the basis for why fossils exist. Fossils occur in sedimentary rocks because these rocks form on the surface of the Earth. The two other basic rocks types, igneous and metamorphic, typically form below the Earth's surface where heat and pressure erase almost all signs of biological remains. Volcanic rocks, a type of igneous rock, cool on the surface, but their heat destroys fossils in almost all situations.

When the Earth formed 4.6 billion years ago, the surface looked nothing like it does at present. No mountains. No seas. No continents. Very little color. (Think of our moon.) Eventually, the planet began to change and cool. A cold, hard crust formed, and dense iron and nickel sank to the planetary center. Between the two layers, a very thick, warm, Play-Doh-like layer developed, called the mantle. Water condensed, oceans formed, life evolved. All within the first billion years.

By this time, the Earth's structure could be compared to a hard-boiled egg consisting of the core (the yolk), the mantle (the egg white), and an outer crust (the eggshell). The crust, like many a hard-boiled egg's shell, is not whole but broken into separate units called plates (because they are very thin features). Unlike a hard-boiled egg's shell, the Earth's outer crust has two varieties: thin, dense oceanic crust and thicker, lighter continental crust. The entire crust of the Earth is relatively cold compared with the mantle layers underneath and is the only layer composed entirely of brittle rock. Circulation of warmer, somewhat mobile areas within the underlying mantle layers forces the cold crust to move around the Earth's surface. Broken at present into seven major and eight minor plates, the crust is constantly moving, driven by the planet's internal heat, which causes the mantle to deform and move the crust—a process known as plate tectonics.

Plate Tectonics

As evolution is the fundamental underlying principle of biology and paleontology, plate tectonics is the basis of geological processes on Earth. The surface of the globe is shaped by weather and long-term climate fluctuations, but the building and destruction of the crust—the rock that makes up the ocean floor and the continents—is the work of plate tectonics.

Most of the Earth's plates are large—for example, those that carry the continents of Africa and Australia (continental crust) and the enormous Pacific Plate (oceanic crust). Some plates are considerably smaller, such as the oceanic Juan de Fuca Plate, which is off the coast from northern California to southern British Columbia.

Plate motion is very slow from our perspective, just 1.5 to 4 inches (3 to 10 cm) per year, and tectonics happens on geological timescales. Plate motion is generated by vast eruptions of basaltic lava along the boundary between two oceanic plates. This new ocean crust pushes each plate away from the boundary, called a divergent plate margin. One local example is the small Juan de Fuca Plate as it moves east at its point of origin on the Pacific seafloor. Where two tectonic plates collide, the boundary is called a convergent plate margin. When tectonic plates converge, the result depends on what the plates are made of.

If an oceanic plate consisting of dense basalt rock collides with a plate of less-dense continental crust, the ocean plate descends and the continental plate rides atop what is called the subducting plate. This is where large-scale geology happens. As the plates collide, the jostling generates earthquakes. Also, as one piece of crust subducts the other,

MODERN PLATE ARRANGEMENT

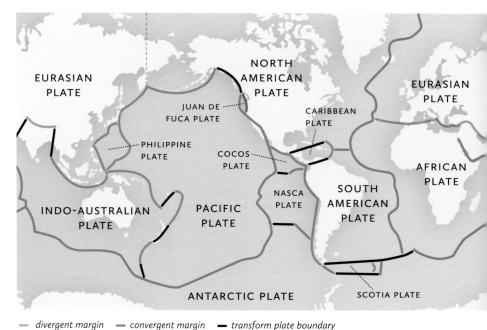

increased pressure and temperature generate magma and volcanoes erupt. Locally, the Cascade volcanoes that stretch from northern California to southern British Columbia are the result of the small oceanic Juan de Fuca Plate pushing east and subducting beneath the huge continental North American Plate.

Another action-packed convergent margin is where two continental plates collide, which is a collision of similar masses. The low-density continental crusts cannot subduct, so the two plates just keep pushing and pushing against each other. During this process, large swaths of the Earth's crust are crushed and folded into mountain chains on the continent, with high peaks and very deep roots below the surface. These forces result in the deep burial of rocks, and this is the location of most metamorphic rocks. Such a collision is responsible for the Himalayas.

Convergent margins don't always involve two continents. Such collisions typically occur between a continent and smaller island chains, such as the Caribbean islands. In these situations, the smaller landmass can be shoved onto the margin of the larger continental plate, where it is called a terrane, thus adding to the leading edge of the plate. This is how the land we now call Washington—as well as the entire West Coast of North America—was built. Tectonic events such as this happened here in the Mesozoic and Cenozoic Eras.

The Rocks

Knowing the differences in rock types, paleontologists can address the most common question they are asked: "How do you know where to look for fossils?" One of the primary ways is by consulting geologic maps, which show different types of rocks and their ages. Often, mapping geologists note the location of fossils based on their field studies, as well as on older reports and maps. For example, if you are looking for dinosaurs, seek out sedimentary rocks deposited on land in the Jurassic and Cretaceous Periods. If you are looking for whale fossils, search for marine sedimentary rocks that are younger than 50 million years old.

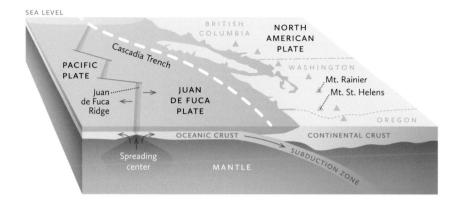

Sedimentary rocks form from the surface weathering and erosion of particles (grains) of all other rocks. Water, wind, or ice then transport and deposit these particles, most of which end up in the oceans. No matter how they are deposited, sedimentary rocks are named by grain size, from the coarse-grained pebbly conglomerates to sandstones to the finest-grained siltstones and mudstones. The finer (smaller) the grain size, the quieter the depositional setting, such as lake beds and the bottom of the sea, and the more chance that organisms survived to be fossilized. Limestone, composed of tiny particles of calcium carbonate that accumulate in shallow tropical seas, is the most fossiliferous sedimentary rock but is very rare in Washington.

Igneous rocks are cooled and crystalized, or solidified, magma. They come in two varieties. Extrusive igneous rocks hardened on the surface after being erupted by volcanoes as lava and ash. Magma that stayed beneath the surface and cooled very slowly is called intrusive igneous rock. Quartz-rich granite, which is relatively low density, is the most common intrusive igneous rock on land, though Washington has very little at the surface. You can see granites around Index, on US Highway 2, the Stevens Pass road, as it transects the Cascade Mountains. Basalt, which covers the ocean floor, is the most abundant extrusive igneous rock. Iron-rich, dark, and very dense, basalt is also widespread across the Columbia Plateau. Large parts of the state are covered by volcanic rock, leaving few areas with good fossil records.

Metamorphic rocks form from the burial and alteration of any type of rock, igneous, sedimentary, or metamorphic. As rocks become more deeply buried under the Earth's surface, such as through the mountain-building forces of plate tectonic collisions, the heat from the layers below combined with the overlying pressure changes the rocks and alters their mineral compositions, which erases all evidence of fossils. These metamorphic rocks reach the surface via subsequent uplift, erosion, and exposure, which is what took place in eastern Washington and the North Cascades region. Washington has a long history of collisional tectonics and a lot of metamorphic rocks—unfortunately for fossil hunters.

GEOLOGIC TIME

As the scientists who study life on Earth through time, paleontologists have to think in four dimensions at once: the height, width, and depth of physical space, as well as the fourth dimension of time across thousands to billions of years. To help understand the immensities of time, scientists over the last two hundred years have developed a sophisticated understanding of geologic time based on two simple ideas.

First, when sedimentary and volcanic rocks are laid down on the Earth's surface, the oldest are at the bottom and the layers get progressively younger up through the pile to its top. This sequence provides a relative measurement of time (timescale names use "upper" for younger rocks and "lower" for older rocks). Sometimes the sequence is disrupted by mountain building, and deciphering these rock sequences can become far more complex.

The second basic principle for determining geologic time is that organisms evolve and become extinct. After an organism or group of organisms becomes extinct, they are never seen again. New biological species then evolve and repopulate the landscape. Some of these species are preserved as fossils, and paleontologists use these changes in individual fossil species or assemblages of co-occurring fossils in the rocks to tell sequences of passing geologic time (timescale names use "late" for younger fossils and "early" for older fossils). For example, dinosaurs lived in the Mesozoic and not before or after; in another example, trilobites evolved about 520 million years ago and became extinct around 250 million years ago, so their fossils do not occur in younger rocks.

The Relative Geologic Timescale

Before the 1960s, geologists relied on a relative geologic timescale developed slowly, mostly within the late eighteenth and early nineteenth centuries, through sequences of rocks and fossils described by European geologists in different countries at different times. Financial exploits initially drove the science, resulting in some timescales proposed by geologists seeking mineral deposits or coal beds or building canals for inland transportation.

In the early 1800s, avocational scientists (before the term *geologist* was used) began to question the relationship between life and the rocks on Earth. In different countries across western Europe, men mapped fossiliferous units and proposed names for sequences of rocks that could be distinguished by their fossil content. (Sadly, very few, if any women, were allowed to be scientists.) At the time, knowledge spread by printed book and pamphlets and by visitors from foreign lands, so the accumulation of a single coherent system advanced over a hundred-year time span.

For example, the Jurassic Period was named in 1799, initially based exclusively on a distinctive limestone unit in Switzerland's Jura Mountains. Subsequent paleontologists studying this distinct limestone then described individual assemblages of ammonite fossils. These fossils were later recognized in other sedimentary rocks across Europe, with species restricted to a short sequence of rocks in each country. In some places, dinosaurs were preserved with ammonites, which gave paleontologists inklings of when dinosaurs lived. Equipped with this information, geologists not only determined that the Jurassic Period in

Europe was a time when global sea levels were high, and warm, shallow seaways frequently flooded most of the continent, but they also had a means to begin to apply the Jurassic time period to other locations around the world.

By 1850 the basic elements of a geologic timescale were established in Europe, with the names of each period and era in place. These were all within the Phanerozoic Eon, and older rocks were simply called Precambrian, with only regional subdivisions based on large-scale geology. (We now know the Phanerozoic Eon is the last part of Earth's history and covers only 12% of geologic time.)

Geologists based most of the periods in the timescale on marine invertebrate fossils because they are common in Europe. On the other hand, they named the Cretaceous Period for the French word for chalk, in reference to the voluminous chalk cliffs on either side of the English Channel. These rocks contain assemblages of distinctive fossils of ammonites and marine reptiles. Geologists also based a few time units, such as the Carboniferous Period and the Paleocene Epoch, on plant fossils. It is still a difficult task, no matter where the rocks are, to integrate land-based rock units with marine fossiliferous units within a single timescale. The smaller the time unit, the more difficult it is.

Each period is further divided into smaller units called epochs, and each of these is divided into stages (or ages) based on the fossils in these rocks. Regionally, the stages are divided into biozones, but correlation of these across geographic space is not always possible. Because organisms around the world are not the same, correlating rocks across oceans—for example, those around the Pacific rim compared with the Mediterranean—presents enormous challenges.

Up until the twentieth century, scientists had no idea how old any of the periods and epochs they had named were; they could only arrange them in order of oldest to youngest—a relative timescale. However, these early geologists understood that the Earth had to be very, very old for the slow processes of geology to shape the entire planet and to have organisms appear and, after some time, disappear. The concept of the vastness of geologic time was key to deciphering the rock sequences and constructing a geologic timescale.

	EON	ERA	PERIOD	EPOCH	TIME
YOUNGER					Today
	Phanerozoic	Cenozoic	Quaternary	Holocene	11.8 thousand years ago
				Pleistocene	2.58 million years ago
			Neogene	Pliocene	5.33 million years ago
				Miocene	23 million years ago
			Paleogene	Oligocene	34 million years ago
				Eocene	56 million years ago
				Paleocene	66 million years ago
		Mesozoic	Cretaceous		145 million years ago
			Jurassic		201 million years ago
			Triassic		252 million years ago
		Paleozoic	Permian		299 million years ago
			Carboniferous		359 million years ago
			Devonian		419 million years ago
			Silurian		444 million years ago
			Ordovician		485 million years ago
			Cambrian		539 million years ago
	Proterozoic	Precambrian			2.5 billion years ago
OLDER	Archean				4 billion years ago
	Hadean				4.54 billion years ago

The Numerical Geologic Timescale

In order to tie together the entire timescale, both temporally and geographically, geologists since the 1960s have also been working on numerical time for the true age for the fossil-based periods, epochs, and stages. Radioactive elements trapped within minerals have been key. Such elements are unstable in nature and change their subatomic structure to become other elements (usually a nonradioactive element), a process misleadingly called decay. Decay occurs at a specific rate for each element, known as its half-life. Geophysicists can analyze the volume of elements within different minerals and employ a formula to determine when, in geologic time, each mineral formed. This is known as radiometric dating.

Probably the best-known radiometric dating uses the rare isotope carbon-14 (written as ^{14}C), but this provides dates that extend back only 55,000 years—thus, it is most useful for archaeologists. Geologists, in contrast, commonly use the radioactive isotopes of uranium (^{238}U and ^{235}U), rubidium (^{87}Rb), potassium (^{40}K), and argon (^{40}Ar). Each of these radioactive elements change over time to known, stable (nonradioactive) isotopes on scales of tens of thousands to billions of years. Although each element has a different rate of change, the rate of decay is absolutely constant for each element and has been physically measured: this is not a guess or an educated estimate.

However, there is a fundamental problem with radiometric dating for paleontology: fossils are found in sedimentary rocks, and almost all radiometric dating techniques require minerals from igneous rocks. Volcanic ash deposits are ideal for finding these rare minerals with radioactive elements because they are geologically instantaneous and they occur on the surface of the Earth with layers of sedimentary rock. Fossils, however, are not preserved in volcanic rock—the heat destroys them—but geologists can obtain numerical ages from overlying or underlying lava or pyroclastic flows to get maximum or minimum ages for the fossils. Most fossiliferous units, though, are not bracketed by lava flows. Nor do they include volcanic ash.

Building the numerical timescale has required a long process of correlations across geographic space, and many geologists are still involved in the endeavor. Over time, the technologies have greatly improved—what once took an entire room of machinery now relies on just a small desktop model—providing greater and greater precision. Unfortunately, the technology is complex and expensive, so very few dates are available for most of the fossiliferous units across the world. Despite this challenge, researchers have established an international chronology based on data from the best localities across the world where fossiliferous units have radiometric dates.

Radiometric dating is not the lone technique used to investigate time in rocks. Paleomagnetism is easier, but no less expensive, to obtain. It relies on the fact that the Earth acts like a global bipolar magnet with north and south poles. As iron-rich sediment particles settle out of the water or air into layers, and as iron-rich minerals in igneous rock crystallize, they align themselves to the Earth's magnetic field. Right now, the north magnetic pole is close to our geographic north pole in the Arctic Ocean, and by convention this is called the positive pole. The negative magnetic pole is close to our south pole in Antarctica.

But over geologic time, the magnetic poles have frequently switched, with the result that the north magnetic pole has been in the southern hemisphere and the south magnetic pole in the northern hemisphere. These switches are not periodic. They can occur approximately every half a million years, as has taken place in the Quaternary, or they can be stable for more than 20 million years, as was the case in the Cretaceous. The magnetic signals in rock show only a positive or a negative alignment for each layer, which necessitates correlation with fossils or radiometric dates to determine the exact placement of the rocks in the timescale. Geologists have put together a timescale of these global shifts in magnetic polarity, and matching reversal patterns can be very helpful if no other method of dating is available.

Many studies continue to refine the geologic timescale by getting more precise radiometric dates, better paleomagnetic signals in the rocks, and better fossil collections from across the globe. The branch of geology concerned with the order and relative position of rock layers and their relationship to the geological timescale is called stratigraphy. The International Commission on Stratigraphy publishes the International Chronostratigraphic Chart online of the entire geologic timescale; it is regularly updated.

GEOLOGIC HISTORY OF WASHINGTON

For the two billion years or so after life evolved, plate tectonics moved the Earth's plates around, creating a curious-looking planet: basically, one supercontinent and one superocean. It is here on the great continent's edge where the future Washington State was born a billion years ago. (When we refer to Washington or any other place in the geologic past, all these locations are the modern incarnations of those long-ago formations.) Somewhere along easternmost Washington and British Columbia, land met the ocean. Across extensive coastal flood-plains and mudflats devoid of plants and animals, rivers deposited vast beds of sediments, now preserved as sandstones, limestones, and siltstones in the far northeastern corner of the state, as well as in much of Idaho, Montana, British Columbia, and Alberta. The very thick rock sequences of marine and nonmarine sediments, now known as the Belt Supergroup, are Washington's oldest rocks. Some of these rocks have evidences of thick bacterial mats of photosynthesizing organisms (slime) that grew in the shallows.

The evidence is limited, but geologists know from the rocks that by around 550 million years ago, fine-grained sediments began to accumulate in this warm, shallow tropical sea that was much like modern Florida. Geologists are not exactly sure of the timing of events between the Belt rocks and these sediments because all rocks of this age do not have enough radiometric dates. The fossilized marine invertebrates preserved from this 520-million-year-old tropical sea are the earliest evidence of animal life in Washington. These fossils are found only along the eastern edge of the state, where they occur in rocky layers known as the Addy Formation, Reeves Limestone, Metaline Formation, and Ledbetter Shale. (A formation is a formal geological definition that refers to a sizable area of specific rocks possessing distinctive characteristics.)

What is now eastern Washington was covered with deep seawater for 50 million years through the Cambrian and into the Ordovician, as global seas progressively rose across the land and the coast moved far inland. As global sea levels fluctuated throughout the Paleozoic, this coastal area was periodically covered by seas or coastal plains. Then, after millions of years of these relatively stable geological conditions, Washington began to undergo a complete transformation. The calm was replaced by the chaos and collision of convergent plate tectonics.

Imagine a typical ocean dotted with reefs, submerged and partially submerged volcanoes, and small continents. Now visualize them on a conveyor belt of moving oceanic crust that sends the entire package east toward the west-moving continent, the North American Plate. As the ocean plate converges with the landmass during subduction, the islands and other bits of land bulldoze onto the continent and get stretched, folded, pinched, and broken into jumbled pieces called terranes. These slivers of new land form a complex puzzle consisting of a disparate array of rocks, each with its own time and place of origin. Another result of subduction is that, as the diving plate descends deep enough to melt and produce magma, the magma subsequently rises and erupts as volcanoes, such as the ones found in modern Washington. Subduction can also produce a different effect: continental growth. In our part of the world, terranes accreted onto the North American continental margin at least three times in our geologic history, adding about 200 miles (322 km) of land to the western edge of North America between the Middle Jurassic and middle Cretaceous.

Only a very few, small fossiliferous units survived these collisions. These include the limestones at Concrete, Skagit County, and outcrops in the North Cascades and San Juan Islands. Other terranes in Washington have other sedimentary rocks, including sandstone, as well as the two other rock types: metamorphic, such as marble and schist, and igneous, such as granite and basalt. As this tumultuous period of accretionary tectonics slowed, ocean water lapped across the continental margin, leaving behind marine invertebrate fossils now exposed in the Methow Valley and on the western slopes of Mount Baker.

In the Late Cretaceous and early Cenozoic, the picture changed again, when no subduction affected our coastline. To the east of the continent's edge lay a very wide coastal plain with large, slowly meandering rivers bringing sediment from Idaho. Unlike in the modern Pacific Northwest, no mountains blocked the rivers' paths, and the landscape was much like today's Texas and Mississippi coastal plains. But just offshore were voluminous eruptions of dense basaltic lava, which occurred over a very short time span, from 50 million to 60 million years ago. Most of the oceanic volcanoes erupted below sea level and close to the coastline, though a few pierced the water's surface and flowed onto the edge of the continent. Out of those underwater volcanic vents came the black basalt of the Crescent Formation, now found in western Washington. The rock is part of a vast volcanic terrane known as Siletzia. Consisting of the basement rocks spread from southern Oregon to southwestern British Columbia, Siletzia was the final terrane to dock at the North American continent.

The origin of the Siletzia basalts has been attributed to a hot spot—a long-lived, deep-seated plume of magma below the Earth's crust. A modern example is the plume that created the Hawaiian Islands: the youngest and most volcanically active islands are the ones atop the plume, off the southeastern end of the Big Island, and the oldest, and hence most eroded, islands or volcanoes lie to the northwest. As the North American Plate moved westward and the hot-spot magma plume stayed stationary within the globe, the plume left behind a path of surface eruptions that geologists can track inland from the Washington coast. By the time of the Miocene, the plume was erupting in spectacular fashion and producing the gigantic Columbia River Basalt Group flows in Washington and eastern Oregon. What began along the Washington coast 50 million years ago now continues in Wyoming today, producing the cataclysmic geology of Yellowstone National Park. For an unknown reason—geologists are searching for the evidence but have yet to find it—the plume appears to have quieted down after 50 million years ago. Around this time, sediments began to accumulate in offshore and coastal plain environments rich with plants and animals. The early Cenozoic coast lay along a north–south line, approximately where modern-day Interstate 5 runs. The sediments deposited in this protean world are now preserved in diverse and widespread strata, with some beds layered atop each other and others interfingering. This complexity of layering has led to a plethora of names, such as the Cowlitz, Chuckanut, Lincoln Creek, Blakeley, and Pysht Formations, all fossil-bearing layers discussed later in the book.

Around 40 million years ago, Washington began to look more like its modern self, with the return of subduction along the Pacific Northwest coastline and the formation of an ancient range of volcanoes. We don't know how high these volcanic peaks rose, but we do know they consisted of thick lava and ash. No matter their elevation, they still succumbed to erosion, eventually becoming more hills than mountains, now forming the roots of the modern Cascade mountain chain.

Subduction also led to the formation of the state's other great mountain range, the Olympics. They are not volcanoes but instead are marine sedimentary rocks that were buried, then pushed upward to their high elevations about 15 million years ago as the coastal part of the state was uplifted and squeezed. Due to the shape of the subduction zone, as it bent westward around Vancouver Island, the stress created a giant horseshoe of rock. On the outer edge are the Siletzia basalts and inside are the early Cenozoic sedimentary rocks, which make up the highest Olympic peaks.

Volcanism also occurred east of the eroded ancient Cascade Mountains from 5 million to 17 million years ago. Instead of molasses-like lava forming steep-sided peaks, the eastern lava was more like warm syrup. This was the hot-spot basaltic lava that covered the Columbia Plateau. Over a span of about 12 million years, dozens of vents generated numerous epic flows of basalt, which eventually covered more than 81,000 square miles (210,000 km²) in Washington, Oregon, and Idaho. In the areas and times where and when the basalt didn't flow, plants and animals thrived, in particular a diverse range of mammals ranging from rodents to rhinoceroses. For the past 5 million years or so, eastern Washington has been relatively calm geologically, with intermittent ice ages.

As anyone who lives in or has visited Washington State knows, the cycles of eruptions of the Cascades that began 45 million years ago have not ended. The state has five modern volcanoes; from north to south, they are Mount Baker, Glacier Peak, Mount Rainier, Mount Adams, and Mount Saint Helens. All are younger than 1 million years old. All but Adams have erupted within the past few hundred years and all are considered to be active.

Nor have other geologic processes stopped. In particular, over the past 2 million years, the state has experienced several periods of extensive glaciation. As happened between Columbia River basalt eruptions, the nonglacial periods had large areas of good habitat for a menagerie of beasts, some of which left their remains. We also still experience earthquakes, landslides, and flooding. Such hazards are part of the challenge but also part of the attraction of living in a state with active geology.

Paleontologists define a fossil as the remains of a living organism preserved in rock. Fossils may be mineralized hard parts called body fossils, carbon remains such as leaf fossils, impressions in the rock, or traces that the organisms left in the sediment (tracks, burrows, and bite marks). And, very rarely, soft parts of organisms, such as pieces of scaly dinosaur skin, are preserved along the surface of a sedimentary layer. Most unusual, though, are small, strange fossils that may or may not be invertebrates, are more than 540 million years old, and do not ever appear again in the fossil record.

Traditionally, fossils are older than 10,000 years, the end of the last Ice Age. More recently, though, researchers have begun to change this definition; because so many organisms have been recovered from deep-ocean sediment cores, anything preserved in rock or sediment now merits definition as a fossil. That characterization still does not cover every specimen that a paleontologist would label as a fossil, such as mammoths and woolly rhinoceroses preserved in the ice in Siberia or the extinct giant birds in New Zealand called moas, which survived until 600 years ago. These are also fossils.

Fossils, however, are rare. Scientists estimate that less than 1% of all species that ever lived on Earth are found in the fossil record. But we will never know, because so few organisms become fossils and so many of these are locked in sedimentary rocks buried deep underground, have been engulfed in hot magma, or have been destroyed by plate tectonics. Although an uncountable number of plants, animals, and bacteria have been part of the story of Earth for 3.5 billion years, the fossil record is spectacularly incomplete.

Paleontologists use the term *index fossils* for plants and animals such as trilobites that evolve and become extinct within a short geological time span and are therefore indicative of that time period. With index fossils, paleontologists use the species' name to define the biozone of time. Because geologists have obtained radiometric dates on fossil-bearing rocks over the last fifty years, they have the ability to divide a period such as the Cambrian into short time slots characterized by unique fossils. Paleontologists like index fossils because anytime they find one, they know that they now have a means to date the rocks where they were found.

Fossils occur across Washington State in mountains, deserts, forests, urban building sites, and beaches. Some are big, such as the mammoth tusk unearthed in downtown Seattle. Some are tiny, including wasps so detailed that you can see their wing veins, and many are small enough to require a microscope in order to study them. Others are so rare, such as the mold of a rhinoceros that is preserved within a lava flow, that their very existence amazes paleontologists, and some are so old—in particular, ancient trilobites and sponges—that they predate dinosaurs by several hundred million years. In a state famed for the diversity of its modern environment, there is also a fascinating range of life-forms stretching back more than a half billion years.

How Fossils Are Formed

Fossilization is a multistep process that begins after an organism dies, settles, and is buried in sediment. Bacterial decay then disintegrates the soft parts, leaving behind hard parts, such as bone, teeth, shell, or wood. The next ingredient is groundwater rich in dissolved chemicals such as calcium carbonate (the second ingredient, after sugar, in Tums), silica oxides, and iron oxides. As these fluids percolate through pore spaces in the sediment, they replace the minerals in the organism's hard parts.

In marine sediments, calcium carbonate, or calcite, is the primary replacement mineral, and it can be hard to distinguish the replacement calcite from the organism's original calcite of the shells. After fossilization, if plate tectonic movement has caused the sedimentary rocks to be uplifted and exposed to weathering, the fossil's hard parts—for example, a shell—can dissolve, leaving behind the sediment, now a rock, that filled the inside of the shell. Known as a mold (the outer impression) and cast (the interior filled part), this type of fossil can preserve the detailed morphology of the organism, such as muscle scars on the inside of a clam or ammonite shell. Some of the state's best-preserved marine shells are found in southwestern Washington.

THE FOSSILIZATION PROCESS

top to bottom: living organism in water; death of organism that drops to sediment surface; decay of soft parts and burial in sediment; fossilization of organism's hard parts as the sediment becomes sedimentary rock

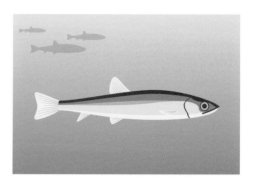

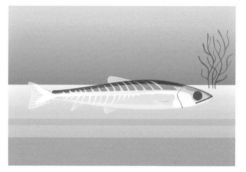

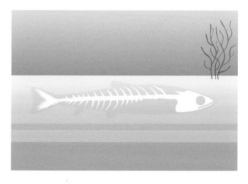

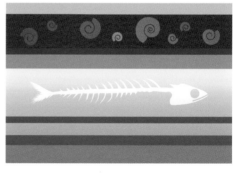

Fossil scallop shell
(*Patinopecten*),
5 inches (12.5 cm) wide

Bone and teeth are the primary body parts of vertebrate animals that fossilize. They are usually mineralized by silicates, including the two most common minerals on the planet's surface, feldspar and quartz, as well as calcite, phosphates, and/or iron oxides. Which mineral depends on the environment, the rock, and the type of groundwater, and different minerals will produce different colors from white to gray, brown to black. In bones, the internal structure resembles a sponge composed of a matrix of hard, mineralized collagen fibers with hollow tubes to carry blood vessels and nerves. During fossilization, minerals replace the bone and fill in the spaces, leaving behind an open, porous pattern in the original three-dimensional form. This is easily identified as fossilized bone. (At the Burke Museum, visitors often bring in chunks of rock, expecting, or hoping, that they have found a fossil bone. Most are simply rock with interesting weathering patterns, and when shown real fossilized bone, the visitors can immediately tell the difference.) In Washington, fossil whale bone from the Pacific coast is heavily mineralized, but some bone and most teeth are barely altered, such as the Ice Age mammals found in southeastern Washington.

Fossilized wood forms in a similar manner to fossil bone, with mineral replacement of the original three-dimensional structure in a process known as petrification. In petrified wood, the minerals frequently replace the wood with little of the original wood fabric surviving, but sometimes the cell structures are preserved too. The best examples of beautifully fossilized wood occur in the Columbia River basin area, in southeastern Washington.

Sponge-like fossil whale bone
in a cemented mudstone,
28 inches (70 cm) wide

Two fossil teeth from
a horse (*Equus*), 8 inches
(20 cm) long

Fossil bald cypress wood
(*Taxodium*), 16 inches (40 cm) long

Fossil leaves
on a slab, 11 inches
(28 cm) wide

In contrast, leaves, seeds, fruit, and occasionally flowers and insects are preserved as impressions within very fine-grained rocks, such as siltstones and shales. Some of these rock layers can be paper-thin and when split reveal the fossil leaf on both the top and bottom rock layer, what paleontologists call part and counterpart. In some localities, the leaf tissue is left as an organic film of carbon and still shows the tiny holes, or stomata, that let gases in and out of the plant. Paleobotanists use these fossilized stomatal patterns to track changing levels of carbon dioxide (CO_2) in the atmosphere over geologic times. Fossil leaves are found across the Puget lowlands, on the lower eastern slopes of the Cascades, and in spectacular lake deposits around the town of Republic, in Ferry County.

A very rare type of preservation occurs when leaves, as well as insects, feathers, and even small frogs and lizards, get trapped and covered with resin, or sap, leaking out of trees. If the resin hardens before burial, it becomes solid, translucent amber, which can preserve interior tissue and even soft tissue as well. There are a few spots where poor-quality amber with pieces of plant material trapped inside has been recovered within Eocene sediments of the Puget Group.

In addition to leaving behind their hard parts, and occasionally soft parts, animals can also leave behind evidence of their behavior, such as tracks and burrows, called trace fossils. Within Washington, there are footprints of giant birds and small invertebrates; burrows of invertebrates including clams, shrimp, and worms; and elongate boreholes in wood of "shipworms," a type of clam that drilled into wood millions of years ago. Their modern relatives have long caused significant damage to pilings, wharves, and piers in Puget Sound. Other trace fossils include eggs, bite marks in leaves, and coprolites (fossilized poop).

Over the past few decades, paleontologists have teamed up with scientists from other disciplines, notably physics, chemistry, and ecology, to use these remnants of past life to tease out interesting stories about behavior of extinct animals. For example, recent studies of fossil tracks have revealed how animals interacted in groups. We frequently don't know which organism left the trace fossil, but by studying behaviors of living organisms we can make educated guesses about the fossil traces and help shed life on the lives of long-extinct flora and fauna.

The present-day distribution of fossil species does not always reflect where an organism actually lived, but tells more about where they died and how they were preserved (called taphonomy). Fossilization is such a rare occurrence, and studying where and how each fossil was preserved can provide a great deal of, but not all, information about how that organism lived many millions of years ago.

How Fossils Are Named

Every type of living organism has a name. Like our own names, these names distinguish one species from all the others so that we know exactly what organism is under discussion. But instead of individual organisms getting a name, in biology this name applies to all members of that same species. Humans are *Homo sapiens*, domestic cats are *Felis catus*, lions are *Panthera leo*. These names originated in Latin with some Greek word roots, and they are printed in italic script to distinguish them from their associated common names. This system of two words—a binomial—for each species was first proposed in 1753 by the Swedish naturalist Carl Linnaeus to eliminate the confusion caused at the time of multiple names for the same organism. The Linnean system works so well that it still prevails and now follows a huge set of international rules.

The first, or generic, name—for example, *Panthera*—is the name of the genus that includes all organisms most closely related, such as tiger, leopard, jaguar, and snow leopard, as well as numerous fossil types. *Panthera leo*, the species name, refers only to modern lions. Using another cat as an example, the youngest (and most famous) saber-toothed cat fossil is named *Smilodon fatalis*. Over the last 40 million years, saber-like canine teeth have evolved many times in different groups of cats, from small to big, and they all belong in different genera. They are cats but not closely related. This naming system also encodes the evolutionary history of the organism (as best the scientists know at the time).

In this book we use the currently correct binomial names of the fossils we discuss, but beware that names are not permanent and can be changed. Because of new and perhaps better-preserved specimens, and with additional knowledge gained over years of work, paleontologists studying a group of fossils may realize that the name may be the wrong one. For example, the giant flightless bird *Gastornis* is now assigned to the generic name used for most of the fossils once called *Diatryma*. In addition, trace fossils such as animal footprints are given generic and specific names so that they too can be distinguished by scientists. Of course, trace fossils do not evolve like plants and animals, but the naming system remains useful for communication purposes.

You've Found a Fossil—Now What Do You Do?

If you find a fossil on your own private land, you can keep it. If you are on land privately owned by someone else, such as many Washington State beaches, land owned by timber companies, and tribal land, you need permission to be on these lands for fossil hunting. Most of the fossiliferous land in western Washington is privately owned. If you have such permission and you find a fossil, you need to contact the landowner, as the fossils are their property.

If you find a fossil on land owned by a public entity, such as city, county, state, or federal property, in Washington State you need permission to remove fossils. These fossils belong to the government agency (and the people of the area). You may get permission to go fossil hunting, but what you find still belongs to that entity. Paleontologists suggest taking photos of the fossils you find and noting the precise location (latitude and longitude) exactly, which can be used to help the agencies' geologists in determining the significance of the find.

Some fossils are very delicate because surface weathering has weakened the surrounding sediment. It is best not to touch them until you have permission to collect. Some fossils, such as petrified wood, are very hardy and can be touched, but they may shatter into many tiny glassy fragments, so be careful.

You are welcome to contact the Burke Museum (which is the official Washington State Natural History Museum), whose paleontologists can try to identify the fossil. They will also help you collect and stabilize larger fossils, such as mammoth bones, if necessary. On the museum's website there is information on how to contact a paleontologist. High-quality photographs with size measurements of your find will go a long way in helping identify it. Burke paleontologists will not take a fossil from you, though they may ask you to donate it if it is scientifically important. The Burke Museum does not buy fossils. Whenever Burke paleontologists collect fossils, they have permits to do so. In addition, there are professional paleontologists at most of our state universities who also identify specimens or who can direct you to the person who knows the most about your particular fossil.

The only formal location open to the public that has been established to collect fossils in Washington State is the Stonerose Interpretive Center in Republic, Ferry County. It is open in the summer, and for a small fee they will provide you with tools and basic instructions on digging. Fossils are sold at rock stores, at rock shows, and online. The vast majority of professional paleontologists do not like this endeavor, as scientifically valuable fossils end up in private collections and are never studied. It also puts a price on fossils, which makes it very difficult for museums to get such specimens to study and to hold for the public to see.

Interest in fossils can become addictive, and you may wonder how you can become a professional paleontologist. You will need an undergraduate degree, preferably in the STEM fields (science, technology, engineering, and math), and a graduate degree in paleontology or related biology or earth sciences. To be a professor or a museum curator, you also need a doctorate. Your field of study can be in any of the many areas touched on by stories in this book—animals, plants, and microbiota.

On the other hand, there are many avocational paleontologists, and you can read about a few of their finds in Washington in this book. Some also write academic papers describing their fossils and the environment in which the organism lived. These are self-taught experts in their areas of interest, and all you need to become one is an inquiring mind and love of the outdoors.

The Quaternary Period

2.58 Million Years Ago to Present

MARKED BY PERIODIC WAXING AND WANING of continental ice sheets and glaciation across all mountain chains, the Quaternary is the last period of the geologic timescale, extending from 10,000 years ago to 2.58 million years ago. Eight major glacial advances have been recorded in Washington in the last 800,000 years, with the greatest spread of ice sheets occurring 20,000 years ago. At the peak of this most recent period, when glaciers moved over northern Europe, Siberia, and Canada, two continental glaciers covered North America: the western Cordilleran Ice Sheet and the eastern Laurentide Ice Sheet.

In Washington, the southern extant of the Cordilleran Ice Sheet, known as the Puget lobe, filled the Puget lowlands, reaching to where Olympia now stands. Around 16,000 years ago, the ice was 4,600 feet (1,400 m) thick at Bellingham, 3,000 feet (914 m) thick in Seattle, and about 1,000 feet (308 m) thick at its snout. East of the Cascades, the ice front stopped about a third of the way south of the Canadian border and left behind characteristic glacial features at the Waterville Plateau around Withrow and Mansfield. These include massive boulders rafted by the ice and a long, low ridge of loose gravel that marked the very edge of the ice before it melted, or retreated, north again.

During the maximum glaciation period, sea levels were around 300 feet (90 m) below modern-day levels along Pacific Northwest coastlines. Western Washington experienced a very cold and dry climate directly south of the ice sheets, much like that of the vast steppes of modern central Asia. The lowered sea level resulted in what is known as the Bering Land Bridge, more than 600 miles (1,000 km) wide, connecting eastern Siberia and Alaska. In the Quaternary, this northern swath of now-inundated land had very little precipitation and was largely ice-free, covered in a very cold grasslands or tundra. When sea levels were low, many animals crossed from Asia into North America for the first time. Archaeologists suggest that humans moved south along the Pacific Northwest coast, possibly in boats, but that evidence is now underwater. Around 10,000 years ago, as the planet warmed, forests started to grow across northern and western Washington again.

The cold Quaternary was the last time that giant animals, such as mammoths, mastodons, and giant ground sloths, lived across Washing-

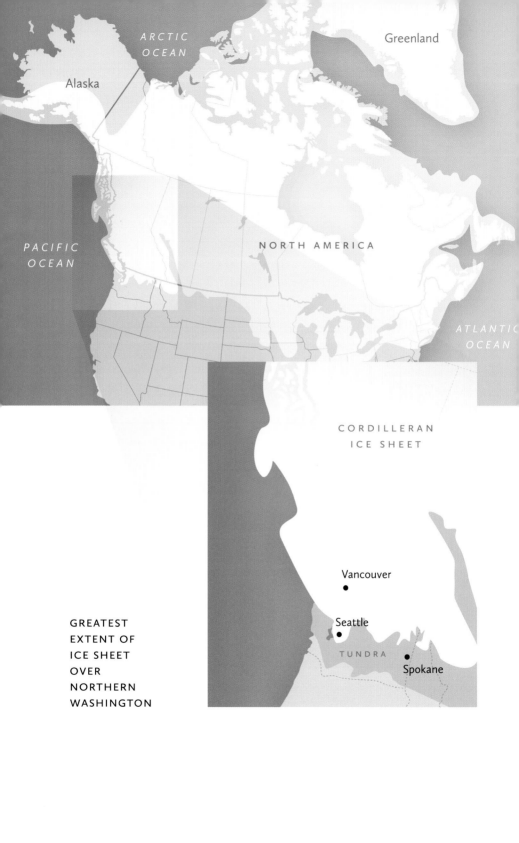

ARCTIC
OCEAN

Greenland

Alaska

PACIFIC
OCEAN

NORTH AMERICA

ATLANTIC
OCEAN

CORDILLERAN
ICE SHEET

Vancouver

Seattle

TUNDRA

Spokane

GREATEST
EXTENT OF
ICE SHEET
OVER
NORTHERN
WASHINGTON

ton. Not only do we focus on these legendary animals and the unique stories associated with their discoveries, we also go to the opposite extreme, exploring some of the smallest fossils, pollen. Together they illustrate the direct connection between our climate, change in global climate, and who inhabits a particular landscape.

Then there's the story of salmon, one of Washington State's most iconic animals. Their tale, much older than the one told by the big mammals and tiny pollen, makes yet another connection between geology and paleontology. Seasonal deposition of sediments on the east side of the Olympic Mountains include many layers with partial skeletons of salmon fossils, which show not only how long salmon spawned in a single lake but also how the lives of those ancient fish are mirrored in the lives of modern salmon. In addition, these fossils exemplify the fact that fossils can be under your feet; you just have to look and be curious.

CHARISMATIC MEGAFAUNA

Mammoths

Thirteen thousand years ago, the northern half of Washington was ripe for change. The Cordilleran Ice Sheet had recently retreated northward, leaving behind a landscape mostly scoured of life. But as has long happened in Earth's dynamic history, plants and animals found their way back. Most would have migrated from the south into the open habitat. Abundant lightweight seeds would have wafted in, germinated, grown, and spread. More seeds would also have been transported by birds and mammals, and as flowers bloomed, insects and arachnids would have descended, seeking out nutrient-rich pollen. Soon, the plants had colonized the recently barren landscape, forming a steppe habitat of mostly open grassy areas scattered with shrubs and trees.

The environment was now ready for the bigger mammals of the Pleistocene—often called charismatic megafauna—to move back into areas they had occupied in previous ice-free periods. They included formidable predators and scavengers, such as saber-toothed cats, dire wolves, and huge bears, as well as massive herbivores: mastodons; giant ground sloths twice as big as the state's biggest land mammal, the moose, which tops out at about 1,000 pounds (450 kg); and bison that were 25% larger than modern ones. All these very large mammals, though, would have paled in comparison to the biggest land animal known to have lived in what are now the boundaries of the state.

Columbian mammoths, *Mammuthus columbi*, weighed up to 11 tons (10 metric tons) and grew to 14 feet (4.2 m) long and as high at the shoulders, with long curved tusks that could measure more than 12 feet (3.7 m) long. As a comparison, the biggest bull elephants from the

Reconstruction of a Columbian mammoth, *Mammuthus columbi*, walking across the grassy tundra in front of glaciers. Open tundra habitat covered much of Washington following the last Ice Age, providing ideal ecosystems for large mammals such as mammoths. Illustration by Julio Lacerda commissioned for the Burke Museum, used with permission.

African savanna, now the largest land animal, weigh around 6 tons and are up to 13 feet (4 m) tall at the shoulder; a few have tusks up to 8 feet (2.4 m) long. (Unfortunately, overhunting and ivory poaching have led not only to fewer elephants but also to a decrease in their average size, notably shorter tusks, and in some places females without tusks at all.) Columbian mammoths are also larger than their well-known relatives, the woolly mammoth, *Mammuthus primigenius*, a species found in Europe, northern Asia, Canada, Alaska, and the northern tier of Midwest states; none have been identified in Washington.

Mammoths, as a group, originated in Africa about 5 million years ago and migrated to Europe, then to eastern Asia. Fossils in Asia show that as the northern hemisphere cooled, mammoths got bigger. They subsequently crossed the Bering Land Bridge into North America about 1.5 million years ago, when sea levels were very low.

Recent advances in technology have allowed scientist to extract and sequence DNA from three mammoth carcasses frozen in the Siberian tundra. Some of the DNA is 1.2 million years old, making it the oldest DNA ever recovered. It further shows these mammoths to be ancestors of both woolly mammoths and the Columbian mammoths, which in places seem to have interbred. All mammoths and mastodons were extinct by the end of the Quaternary, 10,000 years ago.

The great Columbian mammoths would have spent their days walking slowly, feeding, and looking for water, much as modern elephants do. They had very large, straight leg bones, and their feet were covered by thick cushioning pads in order to reduce stress over their long days of trekking. Analysis of dried-out mammoth dung from caves in the American Southwest shows that they ate woody twigs of sagebrush, birch, and spruce, in addition to grasses and sedges. Estimates based on the diet of modern elephants indicate that an adult Columbian mammoth needed a daily intake of more than 400 pounds (180 kg) of food, or about ten times the size of a typical three-foot-long bale of hay. In addition, they would have had to avoid their many predators, including humans; archaeological evidence shows that people were here at this time, but there is some controversy as to whether evidence exists, such as a kill site, for the big beasts being hunted by humans in Washington.

A single molar tooth of *Mammuthus* showing the top washboard-like chewing surface, 12 inches (30 cm) in length; the great depth of the tooth, 11 inches (28 cm) deep; and roots. This specimen and the jaw on page 36 were preserved in a bog in southern Alaska, where tannin in the water colored the bones brown.

Along with elephants, Columbian mammoths are classified within the mammalian order Proboscidea (with trunks) and family Elephantidae. (Mastodons, in contrast, are in their own proboscidean family.) They are all characterized by a very long, mobile nose—the trunk. In order to support the heavy tusks and large molars, as well as the muscles required for both chewing and manipulation of the trunk, mammoth skulls look different from other mammal skulls. They are exceptionally large and high domed. To add strength and maintain lightness, beneath the outer layer of the "face" is a network of bony connections punctuated by abundant open spaces. The brain lies behind this lacy bone structure.

Like modern elephants, which bear the largest mammal molars in the world today, Columbian mammoths had teeth adapted for great food consumption. The molar tooth surface was flat with a series of enamel ridges running across the width, looking like a washboard, perfect machines for grinding plants (and the dirt that was pulled up). Unlike other mammals, elephants have only four molars at one time, two in the upper jaw and two in the lower. When chewing and grinding flattens the ridges, the tooth drops out and is replaced by a new one, which happens six times in their lives. Mammoths had the same method of molar replacement through their lifetime, and molars are the most commonly found mammoth body part.

The lower jaw of *Mammuthus columbi*. Each side of the jaw holds only one very large molar at a time and no incisors. Expanded side flanges 14 inches (35 cm) high held massive chewing muscles. The jaw measures 20 inches (40 cm) long and 18 inches (46 cm) wide at the back.

Dozens of mammoth fossils have been found across Washington, the richest locales being in Jefferson, Clallam, and Island Counties. A partial skeleton with the whole skull and lower jaw was found near Richland in 1978. These fossils came from sediments older than a layer of Mount Saint Helens volcanic ash dated at 12,500 to 13,000 years ago. The bones were excavated and later donated to the Burke Museum, but they were so weathered and fragile that they sat largely untouched on shelves for forty years. Recently, though, 3-D scanning and printing technology of these bones and some from other museums have allowed the Burke Museum to create and mount a complete composite skeleton with approximately 40% original fossil bones and 60% printed silicone copies of bones.

One of the more recent mammoth finds comes from Wenas Creek Valley near Selah, Yakima County. At present, researchers from Central Washington University have assembled an interdisciplinary team of scientists that is incorporating ancient DNA, chipped stone implements, bones of bison and small animals, ground-penetrating radar, luminescence dating, pollen grains, and sediments in order to create a thorough account of the life of the mammoth, as well as the life of the ecosystem and how the different plants and animals related to each other. What makes the Wenas Creek mammoth further exciting is that paleontologists have collected about half the skeleton; most other recent finds around the state are single bones.

The majority of mammoth fossils date from when the last continental ice sheets melted here, 11,000 to 15,000 years ago, but a few older dates on mammoth bones and teeth in Washington extend their range back nearly half a million years. Surprisingly, mammoth remains from at least twenty-four locations have been found in King County, many of which showed up during construction projects in downtown Seattle. To honor the many mammoth bones discovered in Washington, in 1998 Governor Gary Locke signed legislation declaring the Columbian mammoth the state's official fossil.

The first known settlers to find mammoth fossils in the state were the Coplen brothers, in 1876. One excited viewer described the bones as "the grandest discovery of the age to the geological world . . . an animal known to the antiquists as the behemoth." The Coplens had been probing an oozy peat bog on their property along Hangman Creek (now known as Latah Creek, from a Nez Perce word meaning "place to fish"), south of Spokane, when they poked something substantial. After pulling a huge vertebra to the surface, they decided to drain the bog to further their search.

They extracted more than one hundred massive bones from at least six individuals. A single tooth weighed 10 pounds (4.5 kg), the shoulder blade 40 pounds (18 kg), the jawbone 63 pounds (29 kg), and the pelvis more than twice that. But the biggest, topping out at 145 pounds (66 kg) and measuring 10 feet (3.3 m) long, was what the brothers called a "horn." A writer for the *Eugene City Guard* thought the one horn "indicate[s] that the animal was a unicorn." A local schoolteacher realized it was actually a tusk and labeled the beast a mammoth. This was confirmed via photographs of the fossils sent to eminent geologist James Dwight Dana, who in 1841 may have collected what is the earliest known fossil from the state, a block of sandstone from Birch Bay containing leaf fossils and now housed at the American Museum of Natural History, in New York.

To capitalize on their unusual discoveries, the Coplens toured the bones, taking them by steamship down the Columbia River to larger cities. They were not alone. Another set of brothers, the Donahues, who lived nearby, dug into their own bog and pulled up a skull they estimated weighed 800 pounds (362 kg), which they also toured. The bones attracted large crowds and curious comments. They "came from the moon," said one observer. Both sets of brothers finally decided to sell their bones. A composite skeleton incorporating the Coplens' mammoth bones can still be seen in its preeminent location in the Field Museum's Hall of Time (Chicago), whereas the Donahues' skull ended up in the storage basement at the American Museum of Natural History.

Another well-known discovery occurred on February 11, 2014, when construction workers less than three miles south of the Burke Museum found a mammoth tusk and contacted the museum. Working into the night, a team of paleontologists fully exposed and protected the waterlogged soft tusk. They then brought it back to the museum, where it sat drying out very slowly for four years before fossil preparators spent two hundred hours conserving it with glues and preservatives. It is now on display at the Burke, sitting in the lower part of its plaster jacket for stability.

A common misunderstanding in the popular press is the difference between mammoths and mastodons. The two looked similar and belonged in the same order, Proboscidea (animals with trunks), but the American mastodon, *Mammut americanum*, is in its own family, Mammutidae, which has a longer evolutionary history. The earliest members of this family arose some 20 million years ago in Africa and expanded into Europe and Asia. Mastodons migrated from eastern Asia to North America about 10 million years ago when the two continents were connected into a single landmass, millions of years before the Columbian mammoths arrived in the New World.

Mastodons, of the genus *Mammut*, first evolved around 6 million years ago in Asia. Different species migrated across the Bering Land Bridge at different times during the warmest periods of the Pleistocene, some all the way to Central Mexico. Their fossils are most commonly found in the Midwest and northeastern North America, but that is also the location of the greatest number of Ice Age swamps and bogs, areas more suitable for preservation of the bones than the plains or mountains. The most common species across the continent is *Mammut americanum*. Recent studies by paleontologists, however, found that some specimens in Washington, Oregon, Idaho, and California are sufficiently different to warrant a new species—*Mammut pacificus*—and both species coexisted. By 10,500 years ago, at the very end of the most recent Ice Age in North America, all mastodons became extinct.

Mastodons were smaller and stockier than mammoths and had a less domed skull. They also had relatively shorter and heavier legs. Size estimates place the largest bull mastodons at 10 feet (3 m) shoulder height, a weight of 6 tons (5.5 metric tons), and tusks up to 16 feet (5 m) in length. Wear on their tusks shows that they were used for digging, pushing down trees, ripping up bushes, and fighting. Sometimes they broke. In modern African elephants, breakage most commonly

Reconstruction of a mastodon, *Mammut americanum*. These animals were smaller than mammoths, without the domed head. Skin covering and size of the ears are not known but assumed to be similar to those of modern elephants. Illustration by Gabriel Ugueto commissioned for the Burke Museum, used with permission.

occurs during breeding-season fights between males, when tusks get shattered and substantial goring results in broken ribs, shoulder blades, and skulls, as well as death.

The big difference between mastodons and mammoths was the teeth. Unlike the large, flat grinding molars of mammoths and elephants, mastodons had molars with parallel series of large, conical cusps, somewhat similar to pig's teeth in shape. This type of tooth evolved for browsing leaves and twigs. Modern elephants switch seasonally between grazing (eating grasses) and browsing as the nutritional content of grasses changes with the seasons. Geochemical studies of mammoth and mastodon teeth, and dung preserved in dry caves, show that as vegetation patterns changed toward the end of the last Ice Age, both mammal groups ate grasses, tree twigs, branches, and leaves.

Although many artists re-create mastodons covered with thick hair like a woolly mammoth, there is no evidence for pelage in the fossil record. Hairlike red materials found with mastodon fossils and described in early reports have proved to be part of the boggy sediment, including bacterial filaments, collected with the bones. Because mastodons did not live in Siberia, no frozen carcasses have been found in the melting tundra, and we simply do not know about their skin or hair. Without

COMPARATIVE SIZES

🔴 Mammoth

🔵 African elephant

⚪ Mastodon

soft tissue to study, we cannot assess whether mastodons were hairy like mammoths or had skin more like elephants, with minimal body hair.

Hairless or not, mastodons are one of the quintessential Ice Age mammals. Here in Washington, as the last Ice Age began to wane, the continental ice sheets and mountain glaciers melted rapidly and the

The lower jaw of a mastodon, *Mammut americanum*, is 22 inches (56 cm) long and 15 inches (38 cm) wide at the back. The teeth are distinctly different from those of mammoths and are more similar to the cone-shaped teeth of pigs in shape and function.

climate near the ice front was very cold and dry. Considerably lower rainfall forced a major change in the plant ecosystems. Herbivorous animals had to eat whatever was available in their region. In North America, the greatest number of mastodon skeletons have been found south of the Great Lakes area, and this would have been largely a spruce woodland. Mastodons, like modern elephants, most likely walked large distances each day to gather sufficient food, but stayed within an expansive home range. In the end, the climate, and the vegetation, changed more rapidly than the very large mammals could.

Mammut was one of roughly seventy genera of large mammals, or megafauna—broadly defined as more than 100 pounds (45 kg) in weight—that went extinct during the late Quaternary. Many smaller mammals also disappeared, but the impact of losing the large-bodied animals altered entire ecosystems. Like mammoths, mastodon were keystone species, which kept the brush trimmed and helped maintain the diversity of habitats. For example, in southern Africa today, bush areas that no longer have elephants quickly become overgrown with dense shrub thickets, unsuitable for grazers such as antelopes. In the late Quaternary, climate change forced a domino effect of the extinction of the predators that fed on the larger mammals. In North America these included saber-toothed cats, panthers, dire wolves, and giant bears. South America also experienced a major mammalian extinction. Both Europe and Africa suffered less of an end-Quaternary impact, though in modern Africa, where large mammals still persist, extinctions have long been a problem.

The causes of these enormous losses generated fierce scientific debates over the last several decades. There are essentially two polarized camps: those who attribute the extinctions to the rapid climate change at the end of the Quaternary and those who attribute it to habitat destruction and overkilling by humans as they spread across the continents. One of the major stumbling blocks to solving this puzzle has been inadequate and inaccurate radiometric dating of fauna and ecosystems, with grand claims from either side based on unreliable data. Using more refined methods of radiocarbon, or ^{14}C, dating derived from smaller, more pristine samples and with an increased number of specimens, modern researchers have produced a better, tight-knit chronology of the late Quaternary extinction events. In addition, new methods of studying the ecology of herbivores using stable isotope (carbon, oxygen, and nitrogen) analyses of bones, teeth, and even the

A single molar tooth of *Mammut americanum*, 7 inches (18 cm) long and 4 inches (10 cm) deep, is very worn and the enamel covering is cracked and broken, indicating that the tooth sat weathering on the ground surface before it was buried.

food left between the fossils' teeth have produced a more refined reconstruction of how animals lived, what they ate, and how their lives changed after the ice melted and the vegetation changed.

The third major shift in understanding is the analysis of ancient DNA, RNA, and protein from fossil teeth and bones, as well as animal tissue preserved in dry caves and Siberian tundra ice. The techniques of acquiring DNA from late Quaternary animals have been a major breakthrough in understanding biodiversity. For example, DNA analyses of bone fragments retrieved from a cave in France inhabited by humans 45,000 years ago doubled the known number of animal and bird species that lived in the region at the time. Extinction rates may be a lot higher than are currently counted.

Recent studies are also changing the way scientists interpret human damage to large mammal bones—called cut marks. Refined microscopes and innovative use of other high-resolution imaging instruments have frequently shown that the cut marks accumulated not by human butchering but as the fossils were collected, as the animal decayed, or even during intense male rivalry fights.

Washington State has controversial evidence that humans may have killed a mastodon and a bison. In August 1977, Emanuel Manis was excavating a pond on his property near Sequim, on the Olympic Peninsula, when he unearthed two tusks, the larger of which was more than 6 feet (1.8 m) long. A team of researchers eventually excavated the skeleton of a single large old male mastodon. In 2021, a study of mastodon DNA samples from across North America showed that the Manis mastodon belongs in the newly described species, *Mammut pacificus*. This skeleton is of particular interest because of a bone point embedded in one of the mastodon's ribs. The researchers suggested that the point was a human-crafted projectile point and that people had hunted this animal. High-resolution scanning and ^{14}C dating led scientists to conclude that the point had been fashioned by humans 13,800 years ago. The debate on the origin of this bone point continues, with a new study published in 2023 that indicates this point was artificially shaped by human toolmakers but that the blow did not kill the animal.

In 2005 a report surfaced of a discovery at an Orcas Island wetland—called Ayer Pond, after the landowner—of ninety-eight bones of the very large extinct bison *Bison antiquus*. Paleontologists wrote that the bison had lived around the same time as the Manis mastodon and appears to have been butchered. They based their conclusion on cut marks, the character and location of impact marks on the bones, and the bone fracture pattern, as well as the absence of any gnawing or other signs indicating animal predation or scavenging. The researchers hypothesized that the humans who killed the bison could have put him into a frozen pond in order to reduce the aroma that would have attracted scavengers. The carcass then might simply have sunk into the water before anyone returned to collect it.

Not all researchers agree with these conclusions about the connection between early human hunters and the Ayer Pond bison and Manis mastodon. No one questions that humans could have been here at that time, relatively soon after retreat of the ice of the Puget lobe. Instead, paleontologists observe that the location of the point in the Manis mastodon and the possibility that the backhoe used in the excavation could have mashed the point into the animal raise doubts that people killed the animal. Scientists further contend that the Manis mastodon was very old—based on bone chronologies—and that he probably died as a result of a fight with another male or simply old age. The absence of any human-made tools at the Manis site, as well as at Ayer Pond, also puts into question the conclusion of butchering or killing. At present, the most widely accepted evidence for humans in the Puget lowlands comes from a site in Redmond, where archaeologists found hundreds of stone tools and associated remnant cultural objects dated to at least 12,500 years before present.

Although we may not have unequivocal evidence of humans killing the largest megafauna of the Pleistocene in Washington State, people and mastodons—as well as sloths, mammoths, bison, horses, and deer—were here at the same time. It is easy to imagine that such large animals could have provided a key food source—whether scavenged or killed—for the first dwellers of what would become Washington. It must have been both an exciting and a potentially terrifying world to inhabit.

Weighing less than 17 pounds (7.7 kg) and measuring under 30 inches (75 cm) from nose to tail tip, sloths are relatively small tree-dwelling, plant-eating, and leisure-living mammals found in the South and Central America of today. In Puget Sound, Quaternary sloths were a far different beast. They weighed more than a ton and were about as long as a horse, with a barrel-like rib cage, strong thick leg bones, and wide feet. Nor did they live in the trees; the Ice Age ground sloths were much too big to be arboreal. As herbivores, the large body size would have discouraged attacks from predators, such as dire wolves, giant short-faced bears, and saber-toothed cats.

The Puget Sound giant ground sloth is in the genus *Megalonyx*, from the Greek word for "large claw," which references the thick, elongate claws found on the front and back toes. They were probably used to strip leaves from branches and perhaps to dig up tubers, though they may also have aided ground sloths in defending themselves and their young. Complementing their formidable and densely hairy body, ground sloths had a blunt-nosed short skull with a wide, strong lower jaw and peg-like teeth. Unlike other mammal teeth, sloth teeth lacked an outer enamel layer. The soft dentin outer layer wore down rapidly and the teeth continued to grow throughout the sloths' lives.

Pollen studies of sediments contemporaneous with *Megalonyx* and associated Puget lowland species suggest that the animals lived in a cool, moist spruce-dominated habitat, where they browsed in woodlands and river valleys. Fossilized feces preserved in caves from a larger ground sloth confirm that sloths ate leaves and bark from a wide variety of plants.

With their thick leg bones attached to wide-spreading hip bones and a sturdy tail, giant ground sloths could have stood upright on their hind

Reconstruction of a ground sloth, *Megalonyx jeffersonii*, foraging in the tundra grasslands with a herd of bison. Illustration by Julio Lacerda commissioned for the Burke Museum, used with permission.

legs to reach leaves high in trees. There is evidence from Quaternary South American ground sloths that they could stand on their hind feet, but there are no reliable tracks of them walking upright on two feet. In contrast, fossil trackways preserved in dried lake beds in Nevada and New Mexico show ground sloth footprints with all four feet on the ground. Like modern sloths, the animals placed their weight on the outer edge of their feet so that the top of the foot faced outward and the long claws were held off the ground.

The Washington connection to *Megalonyx* began on February 14, 1961, at Seattle-Tacoma International Airport (Sea-Tac), with the first known discovery of giant sloths in the state. Workers excavating a hole for a lighting tower saw bones in the bottom of their work pit. Alerted to the discovery, the Burke Museum sent a paleontologist and an archaeologist to investigate. Although flooding and collapsing walls made the dig difficult, the construction crew and scientists extracted the skeleton, which rested in a peat layer 13 feet (4 m) thick, representing what was once a marshy wetland.

Most of the skeleton was intact except for the skull, which was crushed and mostly missing. Based on the shape of the pelvis, which was 45 inches (114 cm) wide, as well as the limb bones and claws, the Burke paleontologist determined that the bones came from the extinct giant sloth, *Megalonyx jeffersonii*, or Jefferson's ground sloth. Since the initial discovery, other isolated bones and claws of *Megalonyx* have been found in eastern Washington scablands' megaflood deposits and dated at 12,100 years before present. Bones and teeth of a considerably older sloth, *Megalonyx leptostomus*, have also been found in Pliocene sediments in eastern Washington, dated around 4.9 million years old.

In contrast to other large Quaternary mammals, such as mastodons and mammoths, which evolved in Europe or Asia and then traveled east, *Megalonyx* arrived from South America. They came after the isthmus of Panama linked the two continents north to south, joining a northward migration that included opossum, armadillo, and porcupine.

Megalonyx is one of four ground sloth genera that inhabited North America during the Quaternary, 10,000 to 2.58 million years ago. The others were Shasta ground sloth, *Nothrotheriops shastensis*; Harlan's

Skeleton of a ground sloth, *Megalonyx jeffersonii*, standing on hind feet to show its 9-foot (3 m) height. This massive herbivore had a relatively small head, simple elongate teeth, and large claws on each toe to gather leaves and twigs. Courtesy of Triebold Paleo Inc.

ground sloth, *Paramylodon harlani*; and Laurillard's ground sloth, *Eremotherium laurillardi*. Jefferson's ground sloth fossils, which have the widest distribution of the four, have been found across the width of the continent and from southern Mexico to Alaska and Yukon Territory, Canada. The animals are directly related to modern three-toed tree sloths living in Central America.

Megalonyx jeffersonii holds a unique distinction in the annals of paleontology; Vice President Thomas Jefferson gave the animal its generic epithet in 1797. Jefferson, who had a deep passion for natural history, had received several bones of an unknown animal, including the ulna, radius, and claws. The bones had been dug out of a saltpeter (potassium nitrate, a commonly used fertilizer) mine in Greenbriar County, Virginia (now West Virginia). In a March 10 speech in Philadelphia before the American Philosophical Society, Jefferson referred to a "large animal of the clawed kind," which he named *Megalonyx* or "great-claw." He thought the claws belonged to a lion or tiger, though one that was three times larger than a modern variety. Jefferson also suggested that the *Megalonyx* may still have been alive in what was then the vast unexplored part of our continent, and some historians have proposed that Jefferson encouraged Meriwether Lewis and William Clark to look for *Megalonyx* on their expedition.

Jefferson's paper was not formally published until 1799. Curiously, a postscript added by Jefferson notes that his initial interpretation of the *Megalonyx* as a lion was wrong. He had seen a paper written by the great French naturalist Georges Cuvier, which convinced Jefferson that his fossil was akin to a giant tree sloth described by Cuvier. Jefferson's ground sloth received its formal name in 1822, when French anatomist Anselme Demarest named it in honor of Jefferson.

In the summer of 2002, workers constructing a pond on the western side of Orcas Island unearthed another *Megalonyx jeffersonii* specimen. These bones were found in a peat layer that had accumulated in a wetland, lying directly over a shelly marine clay layer deposited by glaciers melting into the newly formed Puget Sound. The Orcas Island sloth remains consist of an entire hind leg and a tooth, found with a few bones of two large bison and a mule deer, the shells of pond snails, and

Foot and ankle bones (bottom row) and two teeth (top left) of *Megalonyx jeffersonii* from Orcas Island, near West Sound. The toe bone with claw (top right) measures 8 inches (20 cm); in life, the claw bone would have been sheathed in a keratin "nail."

pine cones. Other *Megalonyx jeffersonii* bones were found on Orcas Island, but their precise location is not clear. All these sloth bones have been examined for evidence of breakage or cuts from human activity, but none was observed.

So how did these large animals reach Orcas Island? It could have to do with the dynamics of glaciation and sea level change. From about 12,000 to 14,000 years ago, sea levels in Puget Sound and around the world were low because of the immense volume of water tied up in the northern continental ice sheets. In addition, the massive glacier—3,000 feet (914 m) thick in Seattle and 4,600 feet (1,400 m) thick in Bellingham—weighed down the coastal parts of the land. When the ice retreated, or melted, back to the north, the land responded by rising 300 feet (91 m) surprisingly quickly (initially feet per year, then inches per year), over several thousand years—a process called isostatic rebound, similar to when a submerged bath toy rises after the pressure holding it down is released. During the time of low sea level—soon after glacial retreat—islands such as Orcas were much larger, and water channels between them and the mainland were much narrower than at present. Alternatively, sloths could have island-hopped from Anacortes or Guemes Island, which at low sea levels would have been the same peninsula, to Cypress to Blakeley to Orcas Islands.

Modern large mammals can, and do, swim to find new feeding grounds, new territories, and empty spaces; sloths, for example, regularly cross lakes and rivers, albeit rather slowly. Therefore, it makes sense that sloths, bison, and deer could have taken to the water and sought out new habitat in the past. Over the couple of thousand years that the land rebounded, plenty of opportunity existed for large herbivorous mammals to migrate from the mainland and spread across the islands of the Salish Sea. Although we may never know the answer, there is a certain pleasure in simply considering how these large mammals crossed the water.

One reason they may have sought out virgin terrain was the new vegetation emerging on Orcas. In 2016, paleobotanist Estella Leopold published a study of pollen from the late Ice Age lakes on Orcas Island. Her data showed forested areas of pine, spruce, and hemlock interspersed with grass-filled meadows and willow and poplar growing on stream banks, all plants that would have attracted large herbivores.

At present, the Sea-Tac sloth skeleton is on display at the Burke Museum.

Twenty thousand years ago, the area around what is now Battle Ground Lake, south of Mount Saint Helens, was much colder and drier than at present. To the north stretched the Cordilleran Ice Sheet, which was beginning its greatest expanse, ultimately reaching to about 100 miles (161 km) north of the lake. And in the Cascade and Olympic Mountains, alpine glaciers extended far below their modern terminuses. Around the lake, scattered Engelmann spruce, lodgepole pine, and fir grew in a subalpine parkland with grasses, herbs, and plants such as *Artemisia*, though it is unclear if these were sagebrush or smaller species. To find a similar plant assemblage today, you'd have to head up to high elevations on the east side of the Cascades and eastern slopes of the Coast Range in British Columbia. But as is always the situation, the climate was not static, particularly in a glacial landscape.

As the world began to warm up, the Cordilleran Ice Sheet began to retreat from its southern extent. With the retreat of the continental and alpine glaciers came a new climate and new vegetation. By 15,000 years ago, Sitka spruce, mountain hemlock, pine, and shrubby Sitka alder had moved in. This conifer forest, with a high diversity of herbaceous plants and few grasses, thrived in the warmer but still dry environment in southwestern Washington. But this ecosystem, too, would disappear as the climate continued to warm and become wetter. Eventually, this assemblage of trees was replaced by western hemlock and oak-dominated savanna.

How do paleontologists know such detail of what happened so long ago? They study pollen, as well as moss and fern spores that accumulate in the sediment of lake beds. This branch of paleontology, known as palynology, focuses on fossilized pollen grains and spores so tiny—measuring 5 to 200 microns (human hair is about 75 microns wide)—that you need a high-powered light microscope to see them. Few fossils are as small and intricate and yet so informative.

Here in Washington, palynology has been particularly useful to illustrate how the vegetation changed before, during, and since the

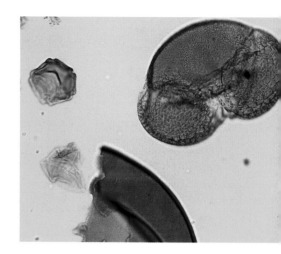

Pollen grains extracted from a bog in the Puget lowlands, viewed through a microscope. The grain on the right is pine pollen, one of the largest pollen grains, 60 microns wide, with two sacs attached that allowed it to float in the air for longer dispersal. The smaller grain on the upper left is alder pollen. Courtesy of Chris Schiller.

last Ice Age. (Pollen can also be extracted from very old rocks and has been the basis of vegetation histories far back in time.) Relatively easy to collect, pollen and spore fossils are extracted from lake-bed sediments via narrow sediment cores a few yards deep. The cores also contain diatoms (silica-shelled algae) and bits of insects that help in reconstructing ancient environments.

In the quiet environment of a lake, sediments accumulate layer by layer through time. For the Battle Ground Lake study, palynologists collected the material they needed by coring directly through the sediment layers with a hand-operated piston corer, 2 inches (5 cm) in diameter and 39 inches (1 m) in length. They worked from an anchored platform in the middle of the lake and recovered more than 49 feet (15 m) of vertical core, which spanned more than 20,000 years. In the lab, the palynologists split the cores open, described the sediments, and took samples at regular intervals down the length of the core. Organic matter in the core, such as wood and charcoal fragments, allowed them to determine the age by using ^{14}C dating techniques. One benefit of working in Washington was that the scientists were able to take advantage of the presence of identifiable volcanic ash layers in the core to add information about the age of individual sedimentary layers within the core. In particular, the Battle Ground Lake cores had ash layers from well-dated ancient Mount Saint Helens eruptions and the massive eruption of Mount Mazama that created Crater Lake in southwestern Oregon 7,600 years ago.

After the samples were in the lab, the hard part of extracting the pollen and spores from the sediment began. It required a lab specifically designed to handle dangerous chemicals used to safely break down the inorganic components of the sediment. The first step involved soaking the subsamples in a series of acid and base washes, from mild potassium hydroxide to the very dangerous hydrofluoric acid. Next, the palynologists isolated, washed, and stained the sediments to increase their visibility, then mounted the residue of pollen and spores on a glass slide and examined it under a microscope at 400–1,000× magnification. To identify pollen grains to family, genus, and sometimes species, palynologists compare them with a "library" of pollen types from living plants.

Counts from each sampled layer of the different types of pollen and spores are usually arrayed on a pollen diagram as number of grains/samples over the length of the core. The radiometric dates and ash layers then allow a time sequence, and the diagram provides a snapshot of the vegetation history of the area around Battle Ground Lake.

As palynologists have continued to study cores samples, they have developed a more detailed picture of the cold, dry period that followed deglaciation. It was during this time of rapid climate change when the mass extinction of large herbivores, including mammoths, mastodons, and giant sloths, along with their top predators, such as saber-toothed cats and dire wolves, occurred across North America. One major cause of the faunal extinction may have been the rapid changes in vegetation in response to climate warming.

At Lake Carpenter, on the northern Kitsap Peninsula, marine diatoms from the lowermost part of the core, dated at 12,800 to 13,600 years ago, show that the area of the modern lake was still part of Puget Sound with marine and brackish water snails and diatoms. From 11,000 to 13,000 years ago, the area around the freshwater lake was dominated first by pines, then by a spruce and pine forest indicating very cold, dry conditions. After 10,500 years ago, a forest of mountain and western hemlock developed, forming a landscape of fewer firs mixed with alder and ferns.

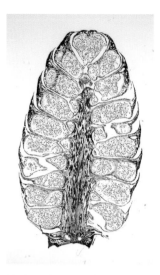

Modern pine cone showing pollen grains (in green). US Geological Survey.

A pollen record has also been studied at Killebrew Lake Fen on Orcas Island in the San Juan Islands by retired University of Washington professor Estella Leopold and her team. The earliest sediment dates to 14,250 years ago, when the island was free of glacial ice. Pollen assemblages from these fen cores suggest a forest composed of pine, mountain hemlock, spruce, and alder, as well as areas of shrubs. Intriguingly, one plant, the seaside juniper, *Juniperus maritima*, was also present and still grows on seaside cliffs on Orcas Island, as well as other San Juan Islands. Around the same time that large herbaceous mammals went extinct and fires became more frequent, about 11,000 years ago, a forest with Douglas fir, western red cedar, oak, and bracken fern began to develop.

Few paleontologists in Washington have had as significant an impact as Estella Leopold. Born in 1927, Estella began studying pollen in the 1950s after earning a PhD in botany at Yale University in 1955 and was hired by the United States Geological Survey soon after. Her groundbreaking research on Paleogene flora in Colorado was some of the earliest and most thorough documentation illustrating how plants respond to climate change. She also played a central role in the creation of Florissant Fossil Beds National Monument, in Colorado, one of a long list of environmentally sensitive places she has helped to protect.

In 1976, she became director of the Quaternary Research Center and professor of botany and forest sciences at the University of Washington. Her work has focused on fossil floras of China, the Rocky Mountains, and the Puget lowlands, including pollen studies that helped reveal a major fault zone, the Seattle Fault zone, which last moved 1,100 years ago. She has also been an influential teacher, mentor to numerous students, and strong proponent of encouraging women in the sciences.

On a hot summer day in 2000, fisheries biologist Jeff Heinis was teaching his new coworker Summer Burdick how to fly-fish. They had driven down an abandoned logging road few people knew near the Skokomish River in Mason County, parked, and hiked into the only accessible location along a narrow, steep-walled canyon. After teaching her the basics, Heinis had headed downstream. "I spent most of my time unhooking my cast," said Burdick, which gave her time to notice that the tan sandstone at her feet was full of fossils. "That's probably why I noticed them. I don't think I caught any fish."

She and Heinis immediately recognized that the fossils were fish and suspected they were salmon, but Heinis was the one who got excited. "He had a lifetime of personal hands-on experience with salmon, and he was sure that this was an extinct species. . . . I would have just left them there. I didn't appreciate their significance," Burdick said. Heinis contacted the landowner, Simpson Timber Company, and biologists with the Skokomish Indian Tribe. "Finding ancient salmon on the Skokomish reinforces the Skokomish people's relationship with salmon. . . . We have always depended on the salmon culturally, economically, and spiritually," said a fisheries manager for the tribe to a local reporter at the time.

When paleontologists from the Burke Museum, University of Washington, and University of Michigan visited the site, they unearthed hundreds of partial skeletons and impressions of salmon scales in the sedimentary layers. None of the collected fossils were whole fish, but there were well-preserved skulls and lower jaws, bodies with backbones, skin and shining scales, and intact tails. The fossils occur within finely laminated layers of lake-bed sediments and adjacent river gravels dating from early in the Quaternary. The fossils are now in the Burke Museum, where fossil preparator Bruce Crowley made a composite of an entire fish, 28 inches (70 cm) long.

Living and fossil salmon belong in the family Salmonidae, along with other familiar fish we eat such as trout, whitefish, and char. *Onco-*

Cast of a complete sockeye salmon, *Oncorhynchus nerka*, made as a composite 28 inches (70 cm) long of four specimens of the salmon skeletons collected from the Ice Age lake beds near Shelton. The hooked jaw of the salmon and the abraded tail indicate that the fish were spawning in the lake. Individual bones and scales can be seen.

rhynchus, the genus of native northern Pacific salmon, includes twelve species, which range from the Gulf of California, Mexico, to Alaska and across to westernmost Russia, Japan, and Korea. Most of these species—known as anadromous fish—spend part of their lives in rivers and lakes, where they hatch and return to spawn and die, and the intervening years in the ocean. Other species remain in fresh water all their lives. Most salmon live from four to seven years.

Researchers initially diagnosed the specimens as a Pacific salmon, in the genus *Oncorhynchus*, based on the large body size 26 to 33 inches (65 to 85 cm) long, small scales, and distinct arrangement of skull bones and teeth. They narrowed it down from the dozen *Oncorhynchus* species that inhabit the northern Pacific Ocean to sockeye salmon, *Oncorhynchus nerka*, by focusing on the straight jawbone, teeth shape and size, and shape of the cheekbone. The more complete skulls, which had the characteristic male hooked jaw, enlarged front teeth (the kype), and abraded tail fins, showed that the specimens were spawning fish. Additional insights came from the growth rings of disc-shaped vertebrae. By counting these annual rings, researchers determined that all the Skokomish River fossil salmon were four-year-old animals. Rings on the individual scales further indicated that the fish spent much of their first year in fresh water, then transitioned to the ocean.

Like sockeye today, these salmon appear to have swum up a river into a lake to spawn and die. The fossils were found near the head of the lake where gravels left by fast-flowing streams had accumulated, usually the kind of places where salmon excavate shallow depressions and lay eggs in redds (egg-depositing sites). However, bouncing rocks and pebbles are not good settings for preserving bodies. The salmon must have made it to the lake, where they died and were subsequently buried in the lake-bottom muds, an excellent place to preserve the fish. (Dead salmon are incredibly important to forest ecosystems because of their rich nutrient load, which often gets transported into the woods by animals that eat and excrete the fish.)

Reconstruction of the fossil sockeye salmon with hooked jaw and red coloring that the fish developed when spawning. Illustration by Gabriel Ugueto commissioned for the Burke Museum, used with permission.

The hundreds of fossil sockeye bones and skin impressions were fossilized within many layers of thinly bedded lake sediments known as varves. These beds, generally less than 1 inch (2.5 cm) thick, typify fine-grained glacial lake deposits. Each varve marks a single year of deposition; thus they can be counted to reveal the time span represented in the rock outcrop. The Skokomish varves show that this site was a salmon spawning ground for about seventy years. (It's amazing to think how rare and special varves are. In contrast with most geological features, which are records of thousands to millions of years, a varve records a single year in the multibillion-year story of Earth.)

Because reliable ^{14}C dating goes back to only 55,000 years ago and these sediments were older, geologists had to use data derived from paleomagnetic signatures in the sediments to narrow down their age. For the past 781,000 years, Earth's north magnetic pole has been close to the north geographic pole. Prior to this time, and back to 900,000 years ago, it was the opposite, called a magnetic reversal. Paleomagnetic investigation of the fossiliferous lake sediments shows that they were deposited during this period, when the north magnetic pole was near the south geographic pole. It was also during a previous ice age, but before the final expansion of ice sheets that extended across northern Washington 20,000 years ago.

Additional insights into the story of these fish come from pollen associated with this lake and other lakes of the same age, which shows that vegetation at the time was typical of very cold, near-glacial conditions. Geologists surmise that some 800,000 years ago, the salmon-rich lake was near but not within the continental glacial front. The salmon could swim from the ocean, probably up an ancestral Chehalis River, into these lakes near Skokomish to spawn.

Older salmon fossils also occur in Washington. Paleontologists have collected isolated lower jaws and vertebrae of the fossil salmon *Oncorhynchus rastrosus* from the lowermost Ringold Formation, north of Richland in Franklin County. The salmon specimens occur with other fish fossils in lake sediments about 6 million years old, during the latest Miocene. Other researchers have also found much more complete specimens of this species in freshwater and marine sedimentary rocks at three localities in Oregon and six in California. They were very large fish, measuring up to 6.5 feet (2 m) long with an estimated weight of 385 pounds (175 kg).

When paleontologists first described and named *Oncorhynchus rastrosus* in 1972, they famously misnamed and illustrated it as a saber-toothed salmon because of the long front fangs developed for spawning. Complete fossil skulls recently discovered in central Oregon, and more extensive fossils of *Oncorhynchus rastrosus* from California, show that these large tusks stuck out sideways and were more like spikes or horns than sabers. The largest collection of *Oncorhynchus rastrosus*, from gravel beds near ancient Turlock Lake in central California, reveal that the extinct spike-toothed salmon, like their modern *Oncorhynchus* relatives, migrated from freshwater lakes and rivers to the sea and returned to the lakes to spawn. It was during this last phase of life that the salmon employed their two spikes, perhaps for fighting and display or possibly to excavate their redds.

Specimens even deeper in the salmon lineage also occur in Washington and southern British Columbia. The oldest salmon in the family Salmonidae is *Eosalmo driftwoodensis*, from the 50-million-year-old volcanic lake beds in Republic, Washington, and near Driftwood Creek, near Smithers, British Columbia. (Paleontologists often use the prefix *eo-*, meaning "dawn," for the name of very early appearances of an animal in the fossil record.) The bones indicate a mix of primitive and more derived features that seem to be intermediate between the salmon and char subfamily (Salmonidae) and the grayling subfamily (Thymallinae). *Eosalmo* was less than 8 inches (20 cm) long. When these ancient lakes existed, there was no river connection between them and the ocean. A variety of sizes in these fossil beds indicates that the juveniles and adults stayed in this lake and, like trout, did not migrate to salt water.

Paleontologists suspect that the 50-million-year-long record of salmon in the Pacific Northwest is not complete. We can only hope that humans and our impact on the planet doesn't bring that venerable run to a premature end.

The Late Cenozoic Era

23 Million Years Ago to 2.58 Million Years Ago

ALTHOUGH THE EARTH EXPERIENCED a general cooling from the Oligocene to the Pleistocene, a short warming period occurred during the middle Miocene around the globe, intensified by geologic events in Washington. Paleontologists have documented that, at 16.9 million years ago, a very rapid and big increase in global atmospheric CO_2 occurred. This coincides with the timing of the Columbia River basalt lava flows that spread across what is now the Columbia Basin in southeastern Washington, eastern Oregon, and Idaho. These extensive floods of iron-rich basalt, which cover more than 81,000 square miles (210,000 km²), erupted from 5 million to 17 million year ago, though 90% of the volume of lava erupted in the first 2.5 million years of activity. These massive, land-based volcanic eruptions, the hottest type of lava on Earth, threw excessive amounts of gases, as steam and both carbon monoxide and carbon dioxide, into the atmosphere (plus other more noxious sulfur and nitrogen gases) that came back down with the rain. The gases would have lofted high into the atmosphere and circulated the whole Earth, adding to the global greenhouse effect. Scientists infer that the heat and moisture would have made eastern Washington, close to the eruption sites, much warmer and wetter than the rest of the region.

As happens in volcanic situations, the great basalt flows were periodic, with variable expanses of time between eruptions. When the noneruptive periods lasted long enough, soils developed on top of the lava flows, which allowed grasslands and forests to develop and expand. Animals then repopulated these ecosystems across the region. During the Miocene, the Cascade volcanoes were also active but with much smaller volumes of magma, which produced a very different type of lava, rich in quartz. Such eruptions would have been far more explosive than the iron-rich basalt flows. By the time the basalt flows began to cease in the earliest Pliocene, the global climate was already cooling down.

In this book we discuss only a handful of fossil sites in eastern Washington, primarily because of the basalt flows. Although the warm, wet environment of the Miocene and early Pliocene was perfect for the lakes, ponds, and rivers where plant and animals flourished, died, and became fossils, many potential fossil sites were probably covered by the lava—but not always. Washington has quite a few sites with an incredible diversity of petrified wood, from trees that lived in mixed deciduous and conifer forests. This wood, composed of colored quartzes, is beautiful enough to be our state gem.

There are other Miocene-Pliocene fossil leaf and insect sites and rare mammal localities in southeastern Washington that we have not included because they are geographically restricted and less studied. However, we do know a lot more about the Miocene-Pliocene ecosystems of Washington that we have included here because of fossil-rich sites in neighboring Oregon and Idaho, where the climate was the same as in Washington but the volcanic eruptions were less extensive.

One of the richest Washington fossil localities of this time is in the younger White Bluffs beds, exposed along the banks of the Columbia River in Franklin County. These beds have produced a wide diversity of early Pliocene mammals, including mastodons, early saber-toothed cats, and modern horses, as well as the fossils we include here—the oldest deer fossils in North America and a rich array of freshwater fish.

Our best-known fossil from this period of eruptions is the Blue Lake rhinoceros. Preserved as a trace fossil within one of the basalt flows, the unique fossil illustrates two aspects of fossils that we stress in the book. The first is that it was found by nonspecialists, two couples out exploring who were not only inquisitive about what they found but who also recognized that they needed to consult experts in order to reveal the scientific importance of their discovery. The second aspect is that fossil formation can be very serendipitous, particularly in a volcanic environment. Other such examples include incinerated log impressions from Mount Rainier and Yellowstone National Parks and the people trapped in their villas in Pompeii, Italy. All these fossils resulted from organisms being enveloped in hot ash.

"It seems that a beauty like the deer's should be rare and hidden. Yet of all the animals who inhabit North America, deer are among the most widespread and abundant," writes Richard Nelson in his wonderful book *Heart and Blood: Living with Deer in America*. The story of these beautiful animals began in what is now Washington with the species *Bretzia pseudalces*, which had large spreading antlers like modern moose and fallow deer. This species is currently among the oldest record of deer on the North American continent, a distinction they share with two other genera, *Eocoileus* and *Odocoileus*; all three reached the continent around 5 million years ago and were descendants of the earliest deer, which evolved in Asia around 20 million years ago.

In the early Pliocene, *Bretzia pseudalces*, or at least the ones that fossilized, lived around a large lake (named Lake Ringold) in what is now the Pasco Basin, in south-central Washington. For nearly 5 million years, between 3 million and 8 million years ago, rivers in the ancient Columbia River valley deposited layer upon layer of sediments in and along floodplains and wetlands, now preserved in the Ringold Formation. Pollen from cores taken from these sedimentary layers record abundant swamp cypress, pine, and juniper trees. The whole region was warm and dry, with grasslands and sagebrush-steppe-type environments and isolated pine woodlands.

Life around the lake and rivers was rich and diverse, comparable to the modern African savanna. There were two species of now-extinct rhinoceros: *Teleoceras*, with particularly short legs and barrel chest and the very large, hornless *Aphelops*. They lived alongside other large grazers and browsers, such as *Megatylopus*, a camel that stood 13 feet (4 m) tall; *Hemiauchenia*, a llama-like smaller camel; and a one-toed horse called *Dinohippus*, which was the ancestor to modern horses. (In this case, the person who created the term *Dinohippus* defined *dino* as "powerful," hence "powerful horse"; with dinosaurs, the prefix *dino* is usually translated as "terrible.") Also dotting the landscape were peccaries and two species of ground sloth, including *Megalonyx leptostomus*, a predecessor of the giant ground sloth found at Sea-Tac

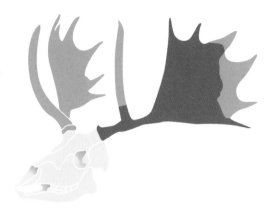

Diagram of a deer skull with the collected antler parts of *Bretzia*. The antlers were spread in a wider angle than those of other deer, more like those on moose skulls. The entire skeleton is the size of a mule deer. Modified from Eric Gustafson, 2015.

(see profile 3), as well as mice, voles, gophers, rabbits, moles, ground squirrels, and beavers. The large browsing animals, such as the deer and ground sloth, indicate that the area was an open woodland, with associated grasslands used by grazing horses and camels.

Such a rich array of large beasts attracted predators of all sizes. Carnivores are represented in the Ringold Formation by *Borophagus*, a large bone-crushing American hyena; *Machairodon*, a small scimitar-toothed cat; and wolverines and weasels. There was also a coyote-type dog and the raccoon-like ringtail cat, *Bassariscus astutus*, the latter of which still live in dry areas from Oregon to southern Mexico.

Like all other deer, *Bretzia* walked on the tips of only two toes (covered by a hoof) on each foot, with all the toe bones elongated to produce a very long, straight leg. Such an adaption is well suited for speed to evade predators. The spectacular *Bretzia* antlers are unique in the state; no other fossilized antlers have ever been found. But they are not unique in the fossil record. All deer, except one species living today, the Chinese water deer, have antlers. In contrast to horns, which grow continuously and aren't shed, antlers grow each year during the breeding season to demonstrate size, strength, and good genes. The biggest ever belonged to Irish elk—a late Quaternary deer that lived from Ireland to Siberia—which had antlers with a spread of 12 feet (3.7 m). Although the shape of antlers has been used to name a number of fossil deer species, antlers from a single species are variable in size and shape—except for caribou, only males have antlers—and are shed in winter and usually not preserved with their skull. In the younger Quaternary fossil setting with numerous species, it's difficult to sort out the individuals and species, though females are typically smaller.

The fossil-rich Ringold Formation rocks extend across the Pasco Basin and Saddle Mountain. At White Bluffs along the Columbia River, upstream from Pasco, the Ringold rocks are in the middle of a sedimentary layer that is more than 1,000 feet (300 m) thick. Similar to many fossil deposits in Washington, the ones that contain *Bretzia* and the associated menagerie were found by a local fossil hunter. Eric Gustafson began collecting in this area when he was in high school in the late 1960s. He was amazed on his first day that he found fossils so easily, mostly of small animal teeth and slivers of tiny bones. Little was known then about this fauna, except that it was Pliocene in age, around 3 million to 5 million years old. In 1970, Gustafson worked with other collectors to assemble bones of all the vertebrate fossils for academic research. Three years later, they described the new deer and other White Bluffs fossils. They derived the new deer's name from *pseudalces*,

Bretzia pseudalces antler, 18 inches (60 cm) in length. This and other antler pieces were collected from White Bluffs, along the Columbia River, Franklin County, in 1970. The pieces of this single crushed (brown) antler were carefully set into (white) plaster of Paris to reconstruct its original size and shape.

Part of the lower jawbone of *Bretzia pseudalces* is 6 inches (15 cm) long
and includes teeth. The molar and premolar teeth define this species as the
earliest deer in North America.

or "false moose," in reference to the antlers, and *Bretzia*, from J Harlen
Bretz, the great geologist who first proposed the idea of the megafloods
that ripped across eastern Washington at the end of the Quaternary.
Gustafson eventually received his PhD and became a college professor—all because of those first collections from White Bluffs.

He also played a key role in describing an equally fascinating collection of fish. Eric Gustafson and fossil fish specialist Gerald Smith
described four separate fish assemblages of different ages found at four
levels in the Ringold Formation. Together, the fish tell a story about the
ancient river systems in the region. Near the base of the formation is
the lone Washington occurrence of *Oncorhynchus rastrosus*, the giant
spike-toothed salmon, previously called the saber-toothed salmon (see
profile 5). These fossils are late Miocene in age, around 6 million to 7
million years old. Because salmon have to return to rivers from the sea
to spawn, their presence indicates that the late Miocene river system
flowed from ancient Lake Ringold all the way to the Pacific Ocean.

But then the salmon disappear. The 5-million-year-old White Bluffs
sedimentary layers instead contain a diversity of freshwater fish including sturgeon, muskellunge, catfish, lake suckerfish, bullhead catfish,
and sunfish. Atop these layers, and 2 million years younger than the
sediments below, the same fossil fish occur, together with pikeminnow
and chub. The notable absence of salmon in the younger rocks show
that Lake Ringold no longer had a connection to the Pacific Ocean in
the Pliocene. The youngest fish fossil section of the Ringold Formation,

located about 15 miles (25 km) north of White Bluffs, in Adams County, is called the Taunton locality. It is another 200,000 years younger. This fish fauna includes all the species found at White Bluffs plus fish typical of ancient lakes in southern Idaho, which had not previously occurred in the Ringold.

The multimillion-year-long changing record of fish in Pasco Basin shows that the region's lake (or lakes) initially had a river outlet that flowed to the Pacific Ocean. But that river connection was severed early in the Pliocene. Still cut off from the ocean in the late Pliocene, eastward-flowing rivers from the Pasco lakes drained into lakes in southern Idaho. Geologists have proposed that this probably marked the beginnings of the Snake River flowing across eastern Washington, a far change from its original route, which had been south and west to California.

This use of fossils illustrates one of their key roles in paleogeography, or study of ancient landforms that are no longer apparent. Fossils can help flesh out details otherwise lost in the geologic record. If no one had found the fossils, and the extra information they provided about fresh- and saltwater fish, then no one would have known the full story of the evolution of the Snake River basin. Similar stories can be told from around the world, each showing how an interdisciplinary approach to science is helping create more-detailed, better-informed, and more-nuanced accounts of past life and past lives.

This multifaceted approach is also illustrative of how the science of paleontology has changed. During the earlier eras of fossil collecting, from the late 1800s into the 1950s, collectors focused primarily on simply naming what they found. In some cases, the race to be the first to publish a name, and thus have that name preserved for posterity, led to errors, such as giving different names to fossils from the same animal based on specimens of different sizes or in different localities. More recently, the focus has been on understanding the plants' or animals' place in an ecosystem and how they related to other species they lived with and other species past and present. In essence, the goal is to help us visualize the extinct plants or animals as they were during their lives.

THE STONE RHINO

Arguably the most unusual fossil ever found in Washington comes from a well-known animal, but one whose relatives are generally more associated with their contemporary home in Africa and Asia. Around 16 million years ago, a rhinoceros lived near what is now Blue Lake in central Washington. At the time, the local environment closely resembled a modern rhino's habitat of a Serengeti-like savanna of mixed woodlands and grasslands, rivers, and lakes.

Roaming the warm, wet landscape of the middle Miocene were a host of grazers and browsers, including three genera of early horses, horned deer, camels that resembled llamas, and four-tusked elephants. Abundant tiny fossil teeth also indicate the presence of shrews, moles, mice, squirrels, and rabbits. They would have been preyed on by carnivores such as small fox-like dogs, ring-tailed raccoons, small slender cats, the occasional small saber-toothed cat, and a large bulky carnivore colloquially called a bear-dog. Most of these mammals are early ancestors of modern mammalian groups, but with more primitive features than the ones we encounter at present. Although this description is based on fossils found in the John Day area of northeastern Oregon, which had an environment more conducive to fossilization, paleontologists believe that the same animals most likely would have been found across eastern Washington.

We know about Washington's long-dead rhinoceros because two couples from Seattle were in eastern Washington looking for petrified wood in July 1935. They were about 200 feet (61 m) above Blue Lake on a narrow ledge of a basalt cliff when they discovered an odd-looking opening. Just wide enough for a single person to enter, the cave was unexpectedly free of the blistered, pocketed surface of a typical basalt flow. Inside, the cave was curved on the bottom with an additional rounded cavity at one end. The cave was about 90 inches (229 cm) long and 36 inches (91 cm) deep in the center. On the ceiling were four circular openings, in two pairs, one at the opposite end from the larger cavity and the other in the middle, or main body, of the cave.

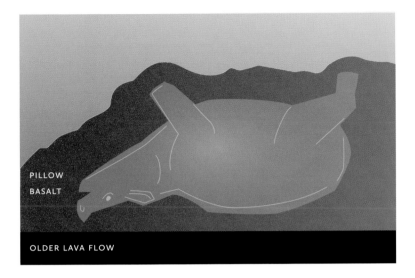

PILLOW BASALT

OLDER LAVA FLOW

Diagram of the cave in a basalt volcanic flow that was formed around a dead rhinoceros, *Menoceras*, 15.8 million years ago. The cave is still accessible after a long and difficult climb up the cliffs from Blue Lake, Grant County.

Paleontologists who subsequently visited the cave wrote that it "has an unmistakable bloated-animal shape . . . [with] features that can be readily identified . . . [as] the legs and head," or at least a carcass that was upside down with legs sticking straight up.

Within the cave, the couples found fossil bone fragments scattered on the ground alongside broken leg bones and a lower jaw and teeth. When the four returned to Seattle, they took the fossils to the University of Washington. Because the geology department lacked a vertebrate paleontologist at that time, the bones ended up with paleontologist George Beck in Ellensburg. He soon ventured to the locality.

Beck wrote that the cave was within a single volcanic flow located within a sequence of many layers of basalt, but he mostly focused on the abundant petrified wood that had first attracted the Seattleites. Noting that the preservation of petrified wood typically occurred in sedimentary rock and not in lava flows, as was the situation above Blue Lake, Beck suggested that dead trees had ended up floating in a lake. Paleontologists surmise that after the rhinoceros died, the body too ended up floating in a lake, upside down, soggy, and bloated. When the basalt flow, which was around 2,000°F (1,000°C), spread into and engulfed the lake, the lava quickly cooled, but the interior molten material remained liquid and plastic-like, preserving floating logs and the shape of the dead rhinoceros. Although the shape of the animal and the mold of the body suggested a rhinoceros, the bone and teeth confirmed the identification. Recently, geologists have shown that the lava that encased the rhinoceros—called the Priest Rapids Member of the Wanapum Formation of the Columbia River basalts—erupted over a very short time 15.8 million years ago.

Beck also sent the Blue Lake lower jaw and teeth to paleontologists at the University of California, Berkeley, who concluded that the fossil was a *Diceratherium*, a genus of extinct rhinoceros. Don Prothero, fossil rhino expert at the Los Angeles Natural History Museum, has indicated that the Blue Lake rhino does not belong to the genus *Diceratherium*, but is more likely *Menoceras barbouri*, which lived across North America at the time. Deposits of vast numbers of *Menoceras* bones have been found at the famous Agate Springs, Nebraska, site, evidence that they lived in large herds.

Adult *Menoceras* had relatively long legs for their squat solid body and were at least 50 inches (127 cm) tall at the shoulder and around 8 feet (2.4 m) long from tail to snout. Modern rhinos top out at about 5,000 pounds (2,268 kg) and up to 14 feet (4.3 m) long. *Menoceras* was the first rhinoceros to have horns (*rhino* means "nose" and *keros* means "horn"), which were side by side near the end of the snout and not the elegant long horn of a modern rhino. As some fossil skeletons have horns and some of the same species do not, paleontologists infer that it's the males with horns and the females without. This sexual

The lower jaw, with some teeth, of the Blue Lake rhino, *Menoceras*, that was found in the cave. These distinct teeth allowed the first paleontologists to identify the cave as a mold of a dead rhinoceros. This specimen shows the standard paleontological museum labeling of two numbers, one designating the locality and one for this individual specimen, as well as a paper tag. University of California Museum of Paleontology.

Reconstruction of a living *Menoceras barbouri*. Art by Mauricio Anton.

dimorphism (two body types) is a common feature in modern grazing and browsing mammals. Sadly, there is a big crack in the rock right across the spot in the fossil mold where the Blue Lake rhino's nose would have been, and we can't be sure whether the rhino was a she or a he.

Rhinos, along with horses and tapirs, belong in the mammalian order of Perissodactyla, meaning with an uneven number of toes. They evolved to walk with hoofed feet across distances to graze on grasses or browse on soft new leaves. Today, perissodactyls walk on a single toe—the third toe—of each foot; the other toes are greatly reduced or absent. But in the Miocene, they were still walking on two or three toes. The earliest known perissodactyl is the first horse fossil, *Eohippus*, which lived around 50 million years ago and stood less than 2 feet (0.6 m) tall at the shoulder. In contrast, the largest perissodactyl is the giant rhinoceros, *Paraceratherium*, which at 18 feet (5.5 m) shoulder height and weighing around 20 tons (the same as four African elephants) was one of the largest land mammals known. Their fossils come from Miocene rocks from Eurasia. They are estimated to have weighed around 40,000 pounds (18,000 kg).

The middle Miocene fauna in Washington lived in a time of chaos and destruction when numerous volcanic vents spewed great flows of oozing black basaltic lava. Between 6 million and 17 million years ago, more than 350 such flows covered more than 81,000 square miles (210,000 km²) of southeastern and central Washington and parts of Idaho, Nevada, and Oregon. Formed from a mantle plume, or hot spot volcanism, the enormous volumes of magma flooded the landscape through hundreds of vents and formed great horizontal layers of rock.

Part of what makes the Columbia River basalts amazing to geologists is that most of the lava erupted within the first 2.5 million years of activity. So much basalt poured out of the feeder vents that it accumulated to a depth of more than 3 miles (4.8 km) in the Pasco Basin. Some of the flows had volumes in excess of 700 cubic miles (2,918 km³), making them some of the largest documented individual flows on Earth. Others were so fluid that they ran all the way to the Pacific Ocean, down what is now the Columbia River Gorge. One flow was so timely that it preserved the carcass of the rhinoceros at Blue Lake, creating one of the world's most unusual fossils.

When the UC Berkeley researchers did their original study, they created an exact replica, or cast, of the interior of the cave at Blue Lake. It eventually ended up in Seattle at the Burke Museum, where it remained in a basement until 1979, when the museum's exhibit designer and fabricator Arn Slettebak used it to re-create a copy of the original cave. He thought it would be the best way to share the tale of the Blue Lake rhino with museum visitors, especially children, who would be able to climb into the replicated belly of the beast. Four hundred and fifty hours of work later, the team completed the new cave, which became a highlight for many visitors. No longer on display, the original mold is still owned by the Burke.

Few fossils in Washington are as beautiful as petrified wood: Rich yellows, reds, and maroons of a sunset. Subtle hues across the spectrum of gray and white. Combine these colors with a repeated pattern of annual growth rings, and you have a recipe for stunning specimens, some of which are dozens of feet long. In a few cases, the fossils resemble a tree so perfectly that a viewer may mistake the fossil for a living specimen.

In part because of the beauty of the fossils, petrified wood became the official Washington State Gem in March 1975. That may seem odd—to designate a fossil as a gem—but the mineral quartz is the main constituent of most fossil wood. Quartz is a very hard, lustrous mineral—silicon dioxide (SiO_2)—which comes in many colors; citrine (yellow) and amethyst (purple) are two well-known gemstones. In petrified wood, the quartz crystals are microscopic varieties known as chalcedony and agate that are colored by traces of iron, magnesium, nickel, or chrome. More than forty other minerals have also been found in petrified wood. One of the strangest occurs in southeastern Utah in petrified trees consisting of uranium ore, which were mined during the Cold War for the nuclear industry.

The formation of petrified wood has long intrigued and confused people. Georgius Agricola, who wrote about geology in the sixteenth century, thought that a stone juice helped form the trees.

One of his contemporaries, theologian Martin Luther, wrote that petrified wood resulted from Noah's flood.

Fossilization of wood requires rapid burial—to prevent decomposition—followed by mineral-rich groundwater penetrating the wood and replacing the wood tissue, sometimes even preserving fine cellular detail. Depending on the level of mineralization, petrified wood can be identified by comparison with living trees. More often, fossilization is imperfect and doesn't give enough information to identify the fossil tree species.

In Washington, Miocene petrified logs are found in many localities across the Columbia Basin. When these trees lived, this region was considerably warmer and wetter. Ginkgo Petrified Forest State Park at Vantage, along the west bank of the Columbia River within Wanapum State Park, is the most accessible. (Fossilized wood has also been found in the Eocene coal deposits of the Puget lowland.) Thousands of logs were preserved here between 15.6 million and 16.1 million years ago, when the forest grew between Columbia River Basalt eruptions. Unfortunately, collectors and raiders have carried away thousands of pounds of specimens. Reminder: It is illegal to collect petrified wood in national parks and any Washington state park.

The Vantage petrified forests occur within the historic lands of the Wanapum people, who made elegant stone tools for trading along the Columbia

Polished slab of fossil *Ginkgo* wood from Vantage, Kittitas County, 11 inches (28 cm) in diameter Although the fossil site within Wanapum State Park is called the Ginkgo Petrified Forest, fossil *Ginkgo* trees are rare. The growth rings of the wood in this slab are clearly visible, and the log was hollow in the middle.

Petrographic thin section of maple wood showing the cellular structure of an unusually well-preserved piece of fossil wood. The width of the whole photograph is 0.12 inch (2.5 mm). Courtesy of T. A. Dillhoff.

River. Scrapers, projectile points, and drills made of petrified wood have been found at archaeological sites across eastern Washington, as well as into the Puget Sound region. These cultural objects date back to at least 8,000 years before present.

Paleontologists consider the Washington fossil woods to be among the richest and most diverse assemblage in the world. The first to study these fossils was paleontologist George Beck, who worked at the Washington State Normal School (now Central Washington University) in Ellensburg. He learned of them in 1930 when a student of his brought him a sack of petrified wood. A year later, while working on his PhD, Beck located specimens in the ground, aided in part by seeing someone carrying a piece of petrified wood down to the road near Vantage.

He and his students eventually collected and analyzed tons of pieces of wood around the Vantage area, many of which ended up in the Burke Museum. Recognizing the importance of the fossils, Beck ensured that the petrified wood site was protected by getting the government to create Ginkgo Petrified Forest State Park in 1935. Originally covering 10 acres, the park now totals 7,470 acres (30 km^2). Today, ginkgo trees grow naturally only in southeastern China, but as you can observe in many neighborhoods, this hardy tree thrives in modern landscaped urban environments.

In a recent study of petrified wood anatomy, paleobotanists Elisabeth Wheeler and T. A. Dillhoff identified woods from thirty-six types of flowering plants and six conifers at Vantage. Most of these can be attributed to modern tree families. Often fossil wood is given a generic name that is exclusively for the fossil wood, to differentiate it from living trees. Very few sites preserve fossilized wood with their leaves or fruit, and paleontologists do not want to make connections that aren't shown by the fossil record. To differentiate the two, paleontologists place the word *oxylon* at the end of the name of a petrified tree. For example, *Hamamelidoxylon* is the wood most similar to the living wood of the witch hazel family, Hamamelidaceae.

Identifying fossil wood is an arduous task that requires slicing the wood with a diamond saw, gluing tiny pieces onto glass slides, then grinding and polishing the specimens by hand to about 0.002 inch (0.04 mm) in thickness, or about one wood cell thick. This allows light to be transmitted through the thin section under a microscope, which can be used to examine the cellular structure of fossilized wood. To make a correct identification requires different orientations of each wood type, cut horizontally, vertically, and at specific angles across the log.

Although Beck noted a high species diversity of fossil woods at the Vantage site, he also found that one wood type made up half the samples he and his students collected. *Piceoxylon* is the wood of spruce and Douglas fir trees. The other common petrified wood is elm (20% in Beck's abundance counts). Other notable varieties include walnut, maple, oak, locust, sweet gum, and witch hazel. In addition, Wheeler and Dillhoff described wood from birch, beech, ash, and plane trees; relatives of flowering plums; and a subtropical mimosa tree. Conifers from Vantage wood also include wood types from the cedar family and yew, as well as the eponymous ginkgo tree. Ironically, the latter are rare.

The Ginkgo Petrified Forest logs represent a forest typical of temperate climates (mixed mesophytic forest) with a high diversity of broadleaf flowering plants. Similar forests, in climates that produce about 50 inches (127 cm) of rain a year, are found today in the Appalachian Mountains in the southeastern United States. The middle Miocene Columbia Basin area was very different from the dry sagebrush environment of Vantage today.

One unusual aspect of most of the fossil wood in Washington is that it is preserved in igneous rock, specifically in Columbia River basalts. The trees fossilized because they grew during one of the many periods following the episodic volcanic eruptions, when soil developed and trees colonized the landscape. When the eruptive cycle restarted, very hot lava flows covered the forests, some of which grew near lakes

Polished slab of fossil oak tree, *Quercus*, from the Ginkgo Petrified Forest,
14 inches (36 cm) in diameter. The colors of fossilized wood are made by trace
inclusion of minerals such as iron, manganese, and sulfur into the quartz
(silica dioxide) crystals that form during petrification.

or swamps. Location was essential to preservation because the water
cooled the lava and prevented it from burning out the entire tree. In
the hills around Yakima, one fossil site has petrified logs still standing
upright.

However, at Vantage geologists have a different interpretation of
tree preservation. They noticed that volcanic pumice—a lightweight
porous rock formed from cooling quartz-rich volcanic explosions, such
as the recent Mount Saint Helens eruption—surrounded the logs. This
and the high proportion of upland trees, such as Douglas fir, mixed
with a typical high diversity of lowland forest trees suggest that the

petrified forest resulted from an eruption-generated mudslide, or lahar. (A similar lahar deposit formed when Mount Saint Helens erupted in 1980.) Then the next Columbia River basalt flow buried this lahar flow and the logs. They were later exposed only when the current Columbia River cut through the rocks.

Paleontologists have identified at least eight significant sites in eastern Washington with petrified wood. Most are located within 50 miles (80 km) of Vantage and range in age from 10.5 million to 16.5 million years old. All of them indicate that what is now the sere, mostly treeless landscape of eastern Washington was formerly covered in lush forests, swamps, and bogs.

One of these fossil localities has a distinct claim. In June 1825, the great botanist David Douglas, who was the first European to collect seeds—about 120 pounds (54 kg), or 3 million seeds—of what is now known as Douglas fir, was traveling up the Columbia River as an employee of the Hudson's Bay Company. At an unknown location along the river, he wrote in his journal that "many large trees in a petrified state are to be seen lying in a horizontal position between the layers of rocks, the ends touching water in many places . . . [some] measure 5 feet in diameter." Along with a slightly earlier journal entry by Douglas's friend, fellow botanist John Scouler, these are the earliest known descriptions of fossils in what is now Washington. Neither Douglas nor Scouler collected a specimen.

The potential honor of collecting the earliest known specimen from Washington goes to James Dwight Dana, a geologist on the United States Exploring Expedition. He collected specimens near Birch Bay in early summer 1841, though it is not clear whether he was on the US or Canadian side of the border. In one of the rocks he collected are fossils from the extinct redwood conifer *Metasequoia occidentalis*.

The Early Cenozoic Era

66 Million Years Ago to 23 Million Years Ago

PALEONTOLOGISTS DESCRIBE the early Cenozoic as a time of unprecedented planetary warmth. The climate from 50 million to 60 million years ago was characterized by very high levels of CO_2 in the atmosphere, substantially higher sea-surface temperatures, and little difference between the climates of the tropics and the temperate zones across the world. In this greenhouse world, even the polar regions were exceptionally warm, with no permanent ice. Fossils found in both the north and south polar regions include palm trees, crocodiles, rhino-like brontotheres, gingko trees, and extensive coal beds. As an example of the warmth, the average temperatures at Ellesmere Island, near northern Greenland, were about 50°F to 54°F (10°C to 12°C), a far cry from today's average temperature of about –4°F (–20°C). The entire Eocene is considered to be one of geologic history's "hothouse" times, with peak temperatures occurring 55 million years ago.

The high temperatures did not last. At the Eocene-Oligocene boundary, 34 million years ago, there was a geologically rapid change that led to very cold currents flowing through all oceans. This cooling of the planet brought ice sheets across Antarctica and to high northern latitudes for the first time since the late Paleozoic some 200 million years before. During these "icehouse" times, the climate progressively cooled through the later Cenozoic with a short and much less dramatic warming period during the middle Miocene.

Few fossil localities were as important to helping scientists understand the huge global climate fluctuations as Washington. On land, early and middle Eocene tropical plants that grew in swamps and along riverbanks became coal deposits along the length of the Puget lowland. This was before the Cascade Mountains rose and huge meandering rivers brought sediment eroded from the granite hills of Idaho, some 450 miles away, to the coastal plains and swamps. With no Cascades, the Okanogan area was the regional highlands, with scatterings of explosive volcanoes.

We have only one fossil vertebrate, a complete turtle, from these river and lake deposits but a wide array of fossil footprints that chronicle life along the muddy edges of the swamps. In the shallow sea to the west of the coast, the high biodiversity of clams, snails, and nautiloids was similar to those from California and the Gulf Coast.

As the Earth was cooling in the late Eocene, the Pacific Northwest was also impacted by the start of the ancient Cascade volcanoes. The rising mountain range soon separated western and eastern Washington, and by the Oligocene the sedimentary signatures in the marine rocks were different from those of the middle Eocene rocks.

These two major changes are recorded in the marine sedimentary layers along the Pacific Northwest coast and by the diversity of fossils of invertebrates and marine mammals and birds. As global climate cooled and warm-water marine animals became locally extinct, we see in the fossil record their replacement by cool-water animals. At the same (geologic) time, the new subducting oceanic plate was active, resulting in the growth of the Cascade Mountains and deep cold-water ocean settings close to the coastline.

We are lucky in Washington because the sediments were buried, turned to rock, and then uplifted above sea level by 25 million years of active plate tectonics. Within these rocks are found a host of unusual animals, particularly fossil whales and dolphins that illustrate the evolutionary split of baleen and modern toothed whales. Other singular fossils in marine sedimentary rocks from this extraordinary time of change that we include in this section are giant flightless birds called plotopterids; *Kolponomos*, a curious animal with both bear and otter characteristics; and a large hippo-like herbivore with teeth unlike most any other animal that ever lived.

Not only were the mammals of the time unusual, but the marine invertebrates also illustrate unusual adaptations. The sediments and fossils indicate a strange deep-water oxygen-poor environment where light hydrocarbon gases (methane and ethane) seeped out of the seafloor and were exploited by a distinct assemblage of clams, snails, and sponges adapted to these transitory sites. Such methane-based communities in the Oligocene are rarely preserved, and Washington has an unusually rich fossil record of such sites, including areas where dead whale carcasses sank into the mud and provided a resource for the mollusks that lived in the low-oxygen ecosystem.

These fossils highlight another point we show throughout the book: that a better understanding of the present helps tell the story of the past. Although paleontologists had long known of these intriguing Eocene and Oligocene fossils, it wasn't until scientists began to study modern hot and cold methane seeps that they began to understand the relationships shown in the past environments.

Reconstruction of a living *Gastornis* leaving its footprints on the mudflats. Illustration by Gabriel Ugueto commissioned for the Burke Museum, used with permission.

Washington State has no dinosaur tracks, but it does have the footprints of birds that could have made most large predators think twice about attacking. These footprints come from the flightless bird *Gastornis giganteus*, a species often called terror birds. The 6- to 10-inch-wide (10- to 25-cm-wide) tracks come from an animal that strode across muddy swamps 54 million year ago. Standing more than 6 feet (2 m) tall, *Gastornis* had a massive skull and beak, a strong, relatively short neck, thick-boned long legs, vestigial short wings, and a body that resembled that of a giant flightless turkey. For many decades, paleontologists depicted *Gastornis* as a predator of small mammals. They originally considered the huge beak suitable for a predatory carnivore, but now paleontologists conclude that *Gastornis* was herbivorous and used their beak for cracking nuts and seeds. The better understanding comes from chemical analyses of *Gastornis* fossil bones showing that the birds fed entirely on plant material.

Gastornis tracks closely resembled carnivorous dinosaur tracks, with three elongated forward-facing toes bearing tiny triangular claws and a deep oval heel pad. (The name honors Gaston Planté, who was the first scientist to find fossil bones of the bird, in 1855 in Paris; they came from sediments that contained broken eggshells, long attributed to *Gastornis* because they are much larger than any other bird's eggs.) As happens in the world of fossilized tracks where no bones exist, paleontologists gave the trackways a specific trace fossil name, *Rivavipes giganteus*, meaning "the footprint of a giant bird on the river." The big birds' fossils are found in both Europe and North America—only in Wyoming and New Mexico. These bones were originally referred to as *Diatryma steini*, but paleontologists have found that this name is not valid. Researchers also assigned to crocodile tracks a trace fossil genus and species, *Anticusuchipes amnis*, which roughly translates as "ancient river crocodile footprint."

Also fossilized with *Gastornis* in the Slide Mountain section of the Chuckanut Formation, near Bellingham, were footprints of other birds and a mammal unlike those in modern Washington. These include heron tracks, each about the same size as a modern great blue heron, or around 4 inches (11 cm) long and wide. Additional tracks have been attributed to turtles, a duck, and other smaller shorebirds that trotted along the water's edge. Unlike the massive *Gastornis*, these birds probably had to be aware of a carnivore that also left behind clawed footprints. These tracks may have been made by a creodont, a predatory mammal about the size of a house cat, which flourished in the Paleocene and Eocene in Europe, Africa, and North America. Creodont fossils have long, narrow skulls similar in shape to a coyote's and carnassial molars, which cut through meat and bone like a pair of sharp scissors. Some modern carnivores also have carnassial teeth; evolution resulted in teeth adapted to do the same task for both groups, although they are not related. Creodonts became extinct in the Miocene.

Also walking in the wet mud was a species that produced abundant large squelch tracks. Most likely made by extinct Eocene grazers— either coryphodontids (large herbivorous mammals) or titanotheres—

Fossil footprints of *Gastornis giganteus* and a heron on a slab cast from rock on Slide Mountain, Whatcom County. The single large *Gastornis* footprint measures 10 inches (25 cm) from heel to tip of middle claw. On this slab, the prints were colored gray for better visibility.

the big round hippo-like tracks show that the animals traveled together. Titanotheres belong in the order Perissodactyla along with rhinoceroses, tapirs, and horses. Bones of these large grazing animals found in Oregon's John Day area and in southern British Columbia confirm that titanotheres were in this region in the early Eocene. Another set of tracks shows fore and hind footprints on either side of a sinuous line made by a tail dragging along the mud through very shallow water. Each footprint is between 2 and 4 inches (5 and 10 cm) long. This was a small crocodile, a testament to the warm and wet early Eocene climate, like central Africa today. They shared the water with an animal whose tracks suggest that it may have been afloat and punting across a mudflat.

What makes these trackways particularly interesting is that paleontologists have found only a single vertebrate body fossil in the Chuckanut rocks. It came from a soft-shelled turtle, or trionychid, which researchers think was the animal responsible for the punting tracks. None of the other animals described above left behind other evidence—either teeth or bone—of their lives. But clearly they lived in and trod across the broad coastal floodplain of meandering rivers and back swamps now preserved in stone.

Trace fossils such as these tracks illustrate another wonderful aspect of geology in that they record a single instant in an animal's life or death. Geology is known as the scientific discipline of millions and billions of years, yet within that deep time of the Earth's story, single moments—a dead rhinoceros getting covered in lava, an insect chewing a leaf, a mammoth defecating, a snail drilling a hole in a clam to get the meat within—are preserved, providing an unexpected and intimate insight into the past.

Further information about the Eocene ecosystem comes from the abundant leaf fossils. Plant remains in the Chuckanut sedimentary rocks near the fossil trackways reflect the subtropical rainforest ecosystem that covered the coastal region of Washington. The most common plants are the huge fronds of the *Sabalites campbelli* palm and *Cyathea pinnata* tree ferns. Other conifers in the fossil flora are the dawn redwood, *Metasequoia*, and swamp cypress, *Glyptostrobus*. (*Metasequoia* still exist in the subtropical areas of China and Vietnam and have been

This large frond of an Eocene fan palm, *Sabalites campbelli*, including the radial fan of leaves, was exposed in a massive rockslide in the Chuckanut Formation, Whatcom County. The geological rock hammer, set on the slab for scale, is 13 inches (33 cm) long. Courtesy of George Mustoe.

transplanted around the world; numerous specimens are found across the state, including at the Seattle Art Museum's Olympic Sculpture Park and the John A. Finch Arboretum in Spokane.) Among the high diversity of flowering plants are fossils leaves and seeds of hydrangea, birch, and plane trees as well as the huge leaves of the fossil sycamore genus *Macginitiea*. (The name honors Harry MacGinitie, who collaborated with Estella Leopold on establishing Florissant Fossil Beds National Monument.)

A notable feature of Chuckanut fossil leaves is the preponderance of large leaves with smooth margins. These types of leaves grow in warm, wet environments, whereas leaves with toothed margins (like those of oak trees) are typical of cooler, drier climates. Paleobotanists studying the leaves have concluded that the average annual temperature for this Chuckanut flora was 61°F to 68°F (16°C to 20°C). This was during the short time known as the Early Eocene Climatic Optimum (EECO), the warmest period of global temperatures in the Cenozoic.

Few people have done more to reveal the diversity of trace fossils in Washington than George Mustoe, a paleontologist now retired from Western Washington University. He has described numerous fossilized tracks, including slabs from the John Henry coal mine near Black Diamond; they contain nine footprints of animals related to modern horses and rhinoceroses and eight tracks of creodonts. In 2009, Mustoe and another collector were exploring Slide Mountain between Deming and Glacier, south of the Nooksack River. A landslide had happened earlier in the year and the pair had gone to investigate. Mustoe was looking at the fossilized heron tracks when his partner pointed out huge bird tracks and asked what they were. Based on previous work he had done, Mustoe immediately recognized that they came from *Gastornis*. He eventually organized a "bird herd" of local fossil collectors to make the arduous climb to the elevation of the footprints on Slide Mountain, where they helped direct a helicopter to carry out a 1,322-pound (600 kg) slab. They also hauled down other trackway-rich slabs for study.

Mustoe's work continues a long tradition of studying trace fossils. The first person known to study them in detail was Edward Hitchcock, a professor of natural history at Amherst College, in western Massachusetts. In 1836, a local doctor had sent Hitchcock a plaster cast of the "tracks of turkeys in relief." Intrigued by the doctor's fossil trackway, Hitchcock created the field he called ornithichnology, the study of bony tracks. The modern term has been simplified to ichnology.

A curious mix of scientist, minister, teetotaler, and vegetarian, Hitchcock amassed thousands of tracks—always wearing his black tie and suit out in the field—one panel of which shows fifty-four beautifully preserved track casts of several species. They are now held by the Beneski Museum of Natural History at Amherst College. Although many of Hitchcock's tracks had been produced by dinosaurs, he could never abide by the idea that God created such monstrous beasts as dinosaurs. Hitchcock always believed that birds produced the tracks. We know that Hitchcock was in a way correct: birds are modern dinosaurs, so it could be argued that Washington actually does have dinosaur tracks.

Coal is a rock. Coal is a fossil. Coal is the fuel that drove industrial revolutions in many countries, as well as a major modern contributor to global air pollution. Often overlooked, coal is widespread across the Puget lowland. These layers of relatively soft, shiny black rock originated in swamps and deltas during the Eocene global hot times and are now part of the rock sequence called the Puget Group, 6,300 feet (1,920 m) thick. The coal-bearing rocks, interlayered with sandstones, occur from British Columbia to Oregon.

The local coal, like all coal, began as very thick layers of dying and decomposing plants that accumulated in swampy areas, which were subsequently buried by overlying plant debris and later preserved by sediment. Plant decomposition led to the formation of peat, the fibrous brown material sold in gardening stores. Over time, as the peat was buried, the heat of the Earth cooked it—deeper burial or proximity to a magma chamber resulted in more heat and greater changes—which led to compaction, water loss, and conversion to coal. Although coal is a fossil fuel, it has little connection to the fossils that people often think of. (Nor does oil come from dinosaurs, another persistent myth, perpetuated in part by Sinclair Oil's use of an *Apatosaurus* as a marketing tool: most oil comes from ocean plankton.) In essence, coal is compacted carbon and hydrocarbons, with small amounts of sulfur and nitrogen.

In contrast to the Cretaceous coal beds of the Rocky Mountains and mostly Carboniferous coal beds of Europe and eastern North America, Washington's coals are much younger. During the Eocene, large meandering rivers flowed out of what is now western Idaho into huge deltaic plains, salt marshes, and swamps, which extended as far east as Kittitas County. The hot, wet ecosystem flourished with a very high diversity of subtropical plants, much like the modern Mississippi River delta. One of the more interesting finds in our local coal beds was the shell of a turtle, 7 inches (18 cm) long, collected in a Roslyn mine and sent to what is now the Smithsonian Institution in Washington, DC, in 1898. Described by O. P. Hay in 1899 and named *Acherontemys heckmani* for mine superintendent Peter Heckman, it is related to modern snapping turtles.

Because of the wide geographic spread of coalfields in the state, geologists have come up with numerous names for the different rock layers that host the coal beds: Chuckanut, Renton, Tiger Mountain, Carbonado, Spiketon, Skookumchuck, and Cowlitz Formations, as well as areas simply labeled as Puget Group Undifferentiated. They share

The top, or carapace, of a fossil turtle shell, *Acherontemys heckmani*, 8 inches (20 cm) long, collected in 1882 from a coal seam in the Roslyn Formation. The white lines drawn on the fossil by the paleontologist who described it in 1899 would not be permitted by the museum today.
National Museum of Natural History

one other characteristic: abundant granitic sand. Because the only source of such grains in the Eocene was granite in Idaho, geologists concluded that rivers carried the sediments out of Idaho and across a flat landscape with no intervening Cascade Mountains. Otherwise, the mountains would have diverted the rivers elsewhere. In addition, there were no mountains on the Olympic Peninsula; the Olympic Mountains begin to form 15 million years ago.

Leaves are by far the most common fossil found in the coal beds, which provided an opportunity for a paleobotanical study of the timing and evolution of the flora. Four sequential time units, or paleobotanical stages—the Franklinian, Fultonian, Ravenian, and Kummerian, each named for a mine or mining town—span the time of deposition from 37 million to 47 million years ago in the Pacific Northwest. The sequences were based on two fundamentals of geology: First, that fossil genera and species are found in restricted time periods as the organisms evolve, appear in the fossil record, and then become extinct, and after extinction, they do not appear again. Paleontological time units are characterized by the entirely restricted occurrence of one or more species, or the overlapping ranges of species that may extend into older or younger rocks. The second fundamental is that during deposition of sediments, the younger layers sit on top of older layers.

These stage delineations came from a suite of characteristic fossil leaf species that do not occur in or are rare in the other units. The oldest unit, the Franklinian, is characterized by early relatives of evergreen beech trees (*Castanopsis*), laurel trees (*Persea*), oleander (*Nerium*), and euphorbia (*Mallatos*). The youngest stage, the Kummerian, which spans the Eocene-Oligocene boundary, includes leaves from camellia, dogwoods (*Cornus*), katsura (*Cecidiphyllum*), and sweet gum (*Liquidambar*) trees. In contrast, conifers, including dawn redwood and swamp cypress, and abundant ferns, tree ferns, and horsetail (*Equisetum*) occur in all the fossil assemblages. None of these trees at present are native to western North America; they are now found in the forests of Korea, southern China, and Japan, as well as in urban gardens in Washington State.

While working on her PhD in botany at the University of Wash-

Fossil leaves from an Eocene relative of dogwood, preserved in an organic-rich dark mudstone layer in the coal-bearing Tukwila Formation of the Puget Group, King County. The rock slab is 9 inches (22 cm) long.

Fern leaf preserved in a fine sandstone layer of the Puget Group rocks, King County. Ferns are common in all the middle Eocene coal-bearing rocks, as they flourished in the tropical climate of the coastal regions. The rock slab is 6 inches (15 cm) long.

ington in the 1980s, Robyn Burnham also studied the Eocene fossil leaves. Her focus, though, was combining biology with geology to better understand the ancient paleoecosystem. When comparing fossil leaves from lakes and ponds, the sides of the river, and flood basins (flatlands away from the main river-channel edges that periodically flood), there is a much higher diversity, and a much better representation of the entire ecosystem, than found in just the flood-basin leaves.

Burnham's work illustrates one of the challenges of geology and paleontology: depositional environments are typically more complicated and nuanced than a simple description implies. In describing the Puget Group rocks, she said, "You could really think floras were changing over time when what was happening was that you were sampling different parts of a large basin including different subenvironments . . . [It] suggests a complicated paleoenvironment, but not really more complicated than a dynamic river system today." In the Puget Group, the most common leaves in the flood-basin deposits include early relatives of flowering shrubs and trees—*Viburnum*, *Alnus*, and a woody vine, *Hypsera*, as well as *Dryophyllum* and *Carya* (walnut and hickory families) and *Bursera* (frankincense)—whereas along the riverbanks, willow tree leaves were most common.

Similar to the geologic story of coal, the economic importance of Puget Sound coal is often overlooked too. William Fraser Tolmie was the first European to mention the local coal. In 1833 the Scotsman, who worked for the Hudson's Bay Company, sailed from London to Fort Vancouver on the Columbia River, then traveled overland to Fort Nisqually, the HBC property on Puget Sound. Near the confluence of the Cowlitz and Toutle Rivers on May 24, Tolmie wrote, "Arrived at coal bed and set the men to work with pickaxe and shovel. The coal did not seem rich, but more of a slatey or carboligeneous nature." Perhaps because of the coal's low quality, the HBC did not try to mine it.

Nor did the Indigenous people of Puget Sound, at least prior to the arrival of Europeans. They had little reason to need coal—wood was abundant, burned cleaner, and smelled better. They also didn't need it to power steam engines, an early primary use of coal. One use, though, may have been as a paint. A nineteenth-century ethnographer wrote of the Twana and Skokomish people that "before the introduction of American paints, black paint was made of coal," though he could have been referring to charcoal.

Commercial mining of coal began in Washington in the 1870s. The earliest mines opened on the east side of Lake Washington along Coal Creek, near what would become the town of Newcastle, and spread to areas with such classic names as Carbonado, Black Diamond, and Cokedale. What started slowly, due to poor transportation, began to boom on March 7, 1877, with the opening of the Seattle and Walla Walla Railroad. Seattle's second train, it introduced a quick, easy, and cheap method of getting coal to the Seattle waterfront. By 1879, Seattle's annual coal exports, almost exclusively to San Francisco, had jumped from 4,918 tons to 132,263 tons. Within another decade, Tacoma had also become an important coal exporter, helping fuel the development of both cities.

The bulk of the state's coal, classified as low-grade bituminous to sub-bituminous (high grade is called anthracite), is loaded with volcanic ash and sulfur that generate airborne pollutants. Washington State's last coal mine, in Centralia, shut down in 2006. What little natural gas we use comes from British Columbia. Most of what remains visible from the coal era are abandoned and dangerous mines; some also leak methane gas. All mines should be avoided and never entered.

In contrast to those hiker hazards is the restoration work accomplished on some of the Centralia coal mines. Deeply mined pits have been filled with the debris left over from extracting the coal, and the landscape has been reshaped so that grasslands, streams, and other wetlands are restored. Around 2,000 acres are now protected wildlife habitats of native plants, bird-filled wetlands, hiking trails, and small agricultural spaces.

"What's in a name? That which we call a rose by any other name would smell as sweet" is one of Shakespeare's classic lines. But what if said rose were made of stone? Or what if that rose wasn't actually a rose but was in the same family as chocolate? It certainly wouldn't smell as sweet and might not be looked on favorably in bearing a false name. But what if that flower was preserved in stone in such exquisite detail that it is easy to imagine that it had died only recently and not 49 million years ago? Few who saw it would complain at the deceit; most would simply exclaim its beauty, despite smelling like a rock.

Such is the case with one of Washington's loveliest fossils, the flower *Florissantia quilchenensis*, found in the fossilized lake beds in the town of Republic, Ferry County. Also included in the fine shales are leaves, fruits, insects, and occasional feathers. Stonerose Interpretive Center is the keeper of these rocks, and the museum holds examples of the great diversity of species that have been found there. The center's symbol is *Florissantia*.

However, fossils were not what initially drew the geologically inclined to the area. As happened across the West, gold was the first attraction. The land, which in 1872 had been set aside as the Colville Indian Reservation, was opened by the federal government on February 21, 1896, for mineral claims. Five years earlier, US government officials had coerced the Colville tribes into ceding their land, alleging that they were not making proper use of it. A federal report noted that "the building of railroads, the creation of towns and cities, [and] the opening of mines . . . would be the greatest blessing." Tell that to people who have lived and thrived in this region for thousands of years.

With the land opened up for exploitation, hundreds of prospectors poured in. The *Seattle Post-Intelligencer* labeled the area the "greatest development and showing ever made by any free-milling camp in the West." The mines, which eventually produced more than 2 million ounces of gold, are now mostly depleted. The gold-bearing quartz veins in the lake beds also produced fossils, which led to the first scientific

This fossil flower, *Florissantia quilchenensis*, the size of a nickel, is an extinct member of the extensive Malvaceae family of plants that also includes hollyhocks, hibiscus, okra, and cotton. Paleobotanists propose that the flowers hung down from a stalk and were pollinated by insects or birds.

description of the fossil leaves in 1929. The author noted the impressive variety of species but considered them to be much younger than Eocene in age, possibly because they were so well preserved.

It was not until the 1970s, though, that interest in the Republic fossils began to grow, primarily through Wes Wehr, and Kirk Johnson, then a teenager and now director of the Smithsonian's National Museum of Natural History. Wehr wrote that "looking at the basalt cliffs and the full moon above them, I began to visualize what the landscape had once been." He was also instrumental in assembling the scientific and local educational facilities at the Stonerose Interpretive Center. Appointed affiliate curator of paleobotany at the Burke Museum, Wehr continued to research, collect, promote, and publish papers about the Republic fossils. He was awarded the Paleontological Society's prestigious Harrell L. Strimple Award in 2003 for his contributions to paleobotany and public outreach.

Between 49 million and 53 million years ago, the land that is now Washington was at about the same latitude as at present. But it was also a time known for the global high temperatures of the Early Eocene Climatic Optimum. In Washington, the EECO resulted in two

LEFT A rare complete fossil pine cone, 3 inches (7.6 cm) long, from the Stonerose sites, Republic, Ferry County. This is one of many exceptionally preserved fossil leaves, flowers, cones, and seeds collected by Wes Wehr and Kirk Johnson in the 1980s.

RIGHT Leaves of the dawn redwood, *Metasequoia glyptostroboides*, are common in the Stonerose sites, Republic, Ferry County. This conifer tree was globally widespread in the Mesozoic and early Cenozoic, but by the end of the last Ice Age its natural distribution was reduced to a mountainous region of southeastern China.

distinct ecosystems. Rocks and plant fossils now preserved in the Puget lowlands indicate a tropical to subtropical climate and a high-diversity mixed conifer and hardwood coastal forest. Higher-elevation areas, such as the Okanogan lakes where the Republic fossils are found, also had high diversity but with many different plants. The Okanogan highlands have been estimated to have been 3,300 to 9,800 feet (1,000 to 3,000 m) above sea level.

The Republic fossils are found in the Klondike Mountain Formation, from sites around Republic and Curlew Lake. The same flora is also found in abundance at sites north of the Canadian border, including at McAbee, Princeton, and Quilchena paleofloral sites, British Columbia. All these fossiliferous localities are very fine-grained lake deposits, interspersed with ash layers produced by explosive rhyolite volcanoes that dotted the landscape. The fossils occur within the very thin shale layers as a single perfect leaf or a jumble of broken leaves and sticks

LEFT Birch leaves from Stonerose sediments look very much like modern leaves of the hardwood birch (*Betula*) or alder trees. Modern birch trees typically grow in temperate climates, and in the warm middle Eocene times they grew in the higher-altitude Okanogan plateau.

RIGHT A complete sycamore leaf, 3 inches (7.6 cm) across, and broken twigs are naturally stained dark brown. Sycamore is the name given to several different trees that have similar-shaped leaves. The most common living genus is *Platanus*, and this 50-million-year-old fossil is *Platanus macginitea*.

scattered across the surface. Because of the very fine volcanic ash and mud at the bottom of the lakes, minimal amounts of oxygen penetrated the sediments, which allowed the leaves to be protected from complete decay. Each is a carbonized brown film of tissue looking surprisingly lifelike in the rock.

Paleontologists have recorded at least 250 plant species at Republic. Conifer fossils include leaves, needles, cones, and pollen of pines, hemlock, fir, spruce, yew, dawn redwood, and coast redwood, as well as the deciduous conifer *Pseudolarix*, or golden larch, and the Japanese umbrella pine. Many of the names show that this middle Eocene flora was similar to the modern conifer trees that now grow in southeastern China, Japan, Vietnam, and Taiwan—and unlike the current conifer forests of the Pacific Northwest. Rare finds in the Republic lake beds include ginkgo and cycad leaves, ancient plants left over from the Mesozoic.

The Republic flora also includes species, in particular flowering plants, more like the forests and woodlands of North America, although all of them are Eocene species. Leaves, seeds, and pollen include birch, beech, hazelnut, sycamore, soapberry, and the Japanese katsura tree, along with many more types of bushes. But the biggest diversity of any group is from the Rosacea family, which includes ancient relatives of apples, cherries, hawthorns, and roses, as well as mulberries, elderberries, Saskatoon berries, and serviceberries. The rocks at Republic are some of the oldest known to include these fossils.

As anyone who tends an orchard or garden knows, plants require pollinators. Based on the variety and number of insects preserved in the Republic shales, these Eocene orchards must have been filled with edibles. Often just bits of wings or hard carapaces, the insect fossils represent nectar- and pollen-feeding groups that had a partial dependency on food from flowers. Flying insects are most common, including wasps, flies, bees, dragonflies, moths, and beetles. In some cases, insect larvae, which would have lived within the waters of the lakes, have also been found. They could have been consumed by the three types of fish preserved in the rock: the earliest salmon in the Pacific Northwest,

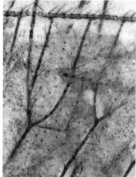

LEFT A crushed mayfly, a species of *Plecia* in the bibionid fly family. Living *Plecia* are commonly called lovebugs. Many whole or partial insect fossils are found in the Stonerose lake beds, including beetles, dragonflies, damselflies, bees, and wasps.

RIGHT Microscopic photograph of a fossil damselfly wing shows the spectacular preservation of individual hairs on the wing veins. Courtesy of Bruce Archibald, University of British Columbia.

Eosalmo; an extinct freshwater mooneye, *Hiodon*; and a trout-perch called *Libotonius*.

Paleobotany has been undergoing a radical change in the past few decades, largely based on Cenozoic fossil sites. Previously, researchers typically assigned names to plant fossils based on the modern genera they resembled, especially the fruits and seeds. If a 50-million-year-old fossil looked like a modern oak leaf, it would be placed in the modern genus *Quercus*. This occurred despite most paleobotanists knowing that the fossils were unlikely to be of the same genus as the ones growing today. Additionally, the fruits and seeds are not often preserved alongside the leaves of the same plant, and these have all been given different names to identify them too.

From the late 1980s, paleobotanists began to move away from this historical practice and at present rarely assign a modern species to a fossilized plant. Instead, they look at features of the plant, such as leaf margins, leaf shapes, and size, in order to arrive at the average annual temperature and rainfall during the time the plants were living. The process of sorting plant fossils into groups based on the characteristic leaf features, without taxonomic names, is called morphotyping, a practice that has been standardized and used extensively to better understand past climates, ecosystems, and changes of floral compositions.

Based on the current practice of morphotyping, paleobotanists have concluded that Okanogan highlands in Washington and adjacent British Columbia (where it is spelled "Okanagan") in the middle of the Eocene had a mean annual temperature of 50°F to 54°F (10°C to 12°C), compared with today's average temperature of 40°F to 50°F (5°C to 10°C). Rainfall is estimated to have been around 39 inches (100 cm) per year. Summers were hot and long, winters very mild.

Getting back to Shakespeare and roses, the paleobotanical revolution in how fossils are named seems to lend credence to the Bard's observation: The name is not important. What is important is the essential element of that 51-million-year-old rose of Republic: its shape, size, and relationship to other plants are what tell the critical story, whether it be or not be a rose.

"The pearly nautilus has fascinated humans for millennia," writes contemporary paleontologist Richard Arnold Davis. "Its precise mathematical spiral and subtle coloration have touched our aesthetic souls. But it is more than that, for the pearly nautilus has titillated our intellects, too." Aristotle was perhaps the first to consider the beautiful shell, but not until the 1700s did the first true scientific descriptions appear. Scientists, though, would need another two centuries to have enough high-quality observations to do more than speculate about these animals that the eminent seventeenth-century British scientist Robert Hooke called "very curious, and indeed very wonderful."

Nautiloids are a group (an entire subclass) of mollusks with multichambered shells that are a sister group to the extinct ammonites and to living octopus and squid. Nautiloids, which first appeared in the fossil record in the Early Ordovician around 480 million years ago, were a very diverse group of large and small predators in the Paleozoic seas. Only two genera are still living today, *Nautilus* and *Allonautilus*, both found in the tropical Indo-Pacific Ocean, off New Guinea, Indonesia, the Philippines, and the islands of Micronesia. Sadly, poaching and sale of the beautiful and exotic shells has severely affected populations, and they have become extinct in some areas. They are now listed as endangered in the international Convention on International Trade of Endangered Species (CITES), and trade is regulated.

Biologist Peter Ward has spent an academic lifetime studying living *Nautilus* to gain insights into their sister group, the ammonites. He and his students have spent countless hours diving with the living animals, studying their biology in aquaria with underwater video systems, and using individual tracking devices. Recently they have documented the plight of rapidly diminishing populations of *Nautilus* due to overharvesting for sale of their shells and from overfishing their entire ecosystem, leaving the animals to starve. In a 2016 paper, Ward and his colleagues wrote, "The future survival of nautiluses is in a race not against time, but against a worldwide demand for their ornamental shells."

Nautiloids and ammonites belong in the class Cephalopoda, or animals whose head (cephal) and mouth are surrounded by a big fleshy foot (pod), which is divided into a number of tentacles or arms, eight in octopus and ten in squid and cuttlefish. These creatures' appendages are outnumbered by the up to ninety very delicate small arms found in *Nautilus pompilius*, known by its common name, the chambered, or pearly, nautilus.

All cephalopods are active swimmers and predators, and the living species have the largest brain in the invertebrate world and complex large eyes. Modern *Nautilus* live at great depths, usually more than 1,000 feet (300 m) deep, and often close to areas where tropical vegetation provides abundant nutrients that wash into the marine abyss. At night, they rise from their daytime haunt in the deep to the surface to feed, using sophisticated olfactory organs near their eyes and on their tentacles to detect the odor of prey, which includes little fish, squid, crabs, and lobsters.

The nautiloid *Aturia angustata* with its preserved pearly shell, 2 inches (5 cm) in diameter. These fossils are abundant in some marine mudstone localities in western Washington that were deposited in deeper water offshore. Nautiloids have very little shell ornamentations, in contrast to members of their sister group, the ammonites.

Washington State has not seen a living nautiloid for 15 million years. Those long-extinct cephalopods are in the genus *Aturia*. A widespread and long-surviving genus—from 10 million to 60 million years ago—they lived in all the oceans, even in the Antarctic in the warm period of the Eocene. Specifically, our species is *Aturia angustata*, which was one of the last-known nautiloids. When it went extinct in the middle Miocene, it was the last living nautiloid species on the eastern side of the Pacific Ocean.

Aturia shells are much more complex than other living mollusk shells, such as those of clams and snails. When cut in half, *Aturia* shells show chambers arrayed in a flat planispiral that grew progressively

Three views of an agatized *Aturia angustata*: LEFT left lateral view; MIDDLE front view showing the siphuncle; RIGHT right lateral view. This fossil has no original shell material, and the interior chambers of the shell are filled with agate, a microcrystalline quartz. This type of preservation provides an excellent view of the internal structure of the shell's curved chamber walls and the hollow tube, or siphuncle, connecting each chamber with the animal that lived in the last chamber. In this fossil, the last chamber, which was the largest, is broken off.

bigger as the animal aged. The last chamber, in which they lived, is the largest. It connects to all the other chambers via a flesh-lined tube, or siphuncle. The tube allows the nautiloid to move up and down by pumping gases through the shell to slowly change its buoyancy in the water.

Like all nautiloids, the shells of *Aturia* are composed of two layers. The inner one is pearly, which is usually what we see in the fossil shells. Some localities in Washington have *Aturia* that retain the pearly layer, whereas others lack any remaining shell; only the impressions of the whorl in the rock have survived. In many cases, the interior of each chamber is filled with calcite or silica crystals that grew into the spaces within the rock, creating a series of sparkling geodes.

Aturia are not the lone nautiloids found in the state. In the 1950s, avocational collectors, working with a professional paleontologist in the Eocene Cowlitz Formation in Lewis County, found two almost complete specimens of another nautiloid, *Cimomia hesperia*. These large rotund fossils are 8 inches (20 cm) long and 5 inches (12 cm) wide in the middle, considerably larger than the *Aturia* from the Cowlitz rocks. *Cimomia* was also globally distributed, but very few have been collected in the Pacific Northwest.

Fossil *Aturia* shells are generally uncommon, but numerous specimens have been found together in a few of the rocks representing deeper-water environments in southwestern Washington and along the north side of the Olympic Peninsula. The concentration of *Aturia* in the deeper-water deposits suggests that they had lives similar to modern *Nautilus*, which live offshore and away from shallow-water waves and predators. Some *Aturia* in Washington, however, occur in shallow-water sediments and were most likely washed toward the shore by winds and tides. Most of these are found in cemented mud spheres, or concretions, which helps preserve the delicate shells of *Aturia*, as well other animals, such as whole-body crabs.

Another group of fossils that are usually preserved only in deeper-water settings are crabs, lobsters, and shrimps, collectively called decapods (meaning "ten feet"). This huge order of animals within the phylum Arthropoda, which also includes insects, spiders, and centipedes, is characterized by a many-jointed hard outer skeleton. When a crab dies, the exoskeleton falls apart very quickly at the joints, allowing waves and scavengers to rapidly break up the pieces. However, if the carcass settles into the soft mud of deeper water, where oxygen is scarce, it has a good chance of becoming a fossil. Within this low-oxygen mud, bacteria are still at work and the rotting flesh changes the chemistry of the mud surrounding the dead crab. This allows calcium carbonate to precipitate and cement the mud in a concretion surrounding the animal's body, though typically a concretion of an arthropod does not include a complete specimen.

All the fossil crabs, shrimps, and lobsters we find in Washington's coastal rocks are preserved this way. We also find nautiloids, clams, and snails in these cement "cannonballs." Be warned, though: if you are out collecting concretions on a beach or in a riverbed, you will probably crack open a hundred before you find a fossil worth keeping.

The diversity of marine decapods from Washington's Cenozoic rocks is impressive, though few species are common and most are very rare.

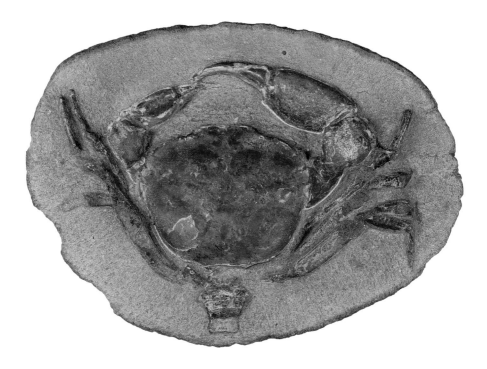

A xanthid or mud crab, *Branchioplax washingtoniana*, preserved inside a concretion 5 inches (12 cm) wide. This fossil was carefully prepared out of the rock to show the details of its features. Preservation of complete crabs is rare.

The cold, deep seafloor environment preserved in the late Eocene through Oligocene sediments, which is typical of fossiliferous rocks in western Washington, produces the most fossils. Most of these are in extinct genera and belong in uncommon families of crabs. A few, though, come from the same families as the living giant Japanese spider crab and the well-known Dungeness crab, along with box and frog crabs, hermit crabs, squat lobsters, and ghost shrimps.

In western Washington, many of the most spectacular crab, shrimp, and lobster fossils were found by Ross and Marian Berglund. Over the years, they collected thousands of concretions, somehow spotting which to try to break open and which to leave on the beach. Ross, born in Seattle in 1913, had a geology degree from the University of Washington and then became a Boeing engineer. When he retired, he was bored and began to collect fossils. He and Marian taught themselves to be decapod-specialist collectors and formed a long-lasting and high-yielding field partnership with Jim and Gail Goedert. Many of the fossils the Berglunds collected were new species or even taxa from new families.

Of critical importance to research paleontologists, Ross and Marian were extremely careful to label every fossil and precisely note its locality. Ross usually prepared the fossil from the rock-hard concretions in his basement workroom, then engaged professional paleontologists from across the country to study them. When he collected new or unusual specimens, he sent the fossils to decapod paleontologists and was frequently coauthor on the publications. In 1994 he received the Harrell A. Strimple Award from the Paleontological Society that honors contributions to the science by avocational paleontologists. Rodney Feldmann, decapod specialist from Kent State University, said, "Ross was one of those rare, passionate people who pursued a hobby with the most careful scientific records and delighted in sharing his discoveries with others." Before he died in 2011, Ross donated much of his superb collection of prepared specimens, and many boxes of unopened concretions, to the Burke Museum, as well as the Smithsonian Institution and a few other West Coast museums.

Acharax dalli is related to the modern awning clam that notably lacks a digestive system. Their nutrients come from chemosynthetic bacteria that live in the clams' gills. This fossil is 2.5 inches (6 cm) long; a very few weathered specimens from the same locality are twice as long.

This fossil of a commonly collected clam, *Lucinoma hannibali*, from the Pysht Formation, Clallam County, is 1.5 inches (4 cm) wide. Living lucine clams host chemosymbiotic bacteria in their gills that oxidize sulfur and other chemicals, providing food for the clams and allowing them to live in low-oxygen deep-sea muds and cold methane seep sites.

The cleft clam, *Conchocele bathyaulax*, 0.52 inch (1.3 cm) long, is another clam species typical of cold methane seep sites in western Washington. These clams are abundant in ancient methane seep sites around the margin of the North Pacific.

One of the classic lessons learned early in many people's schooling is that life on Earth depends on photosynthesis, the process by which plants make energy from sunshine. We are told that without photo-synthesis, life on Earth would not be possible. Several species of clam apparently didn't get this message.

For example, *Acharax dalli* (previously named *Solemya dalli*)—found in Eocene and Oligocene marine rocks across western Washington and northern Oregon—needed no sunlight or even much oxygen. Instead, like living *Acharax* (awning clams), they must have relied on bacteria within their gill tissues, which could manufacture food from hydrogen sulfide oxidation, a process called chemosynthesis. The bacteria pass the food energy directly to the clams that, in return, provide protection for the bacteria. As in all symbiotic relationships, these interactions between the bacteria and the living clam were beneficial to both. In Pacific Northwest fossil sites, *Acharax dalli* is found with a suite of other chemosymbiotic clams, species of the still-living clam genera: the vent clam *Calyptogena*; a cleft clam, *Conchocele*, previously in the genus *Thyasira*; and a lucine clam, *Lucinoma*; as well as deep-sea mus-sels, *Bathymodiolus*.

Unique among this cohort, *Acharax* belongs in the family Solemy-idae, which are distinguished by enormous gills (compared to any other clam) that fill the shell space. They also lack a functional gut and have no other means to obtain food beyond their chemosynthetic bacte-ria. Biologists initially considered the family to be very primitive but now know that the gill tissue harbors the chemosynthesizing bacteria. *Acharax*, *Lucinoma*, and *Conchocele*, which live in burrows within the hydrogen sulfide zones, rely on streams or bubbles of methane and ethane within the sediment—called a seep—bringing the sulfides up to the surface from the deep. *Conchocele* is further aided by using its foot to seek out hydrogen sulfide pockets and, remarkably, can extend the appendage thirty times its usual length to probe the mud for sul-fides around its burrow. All the invertebrates living within seeps carry symbiotic bacteria, and most of them also gather food in the regular way of pulling in detritus-laden water into their shells.

These active cold hydrocarbon seeps and their very unusual animal communities are found in many places around the world today, but particularly around margins of the Pacific Ocean. The underwater canyons off the coasts of Grays Harbor, Washington; Astoria, Oregon; and Vancouver Island, British Columbia, have numerous modern seep sites consisting of thick marine deposits, which accumulated from weathering and erosion of the continental margins. All contain buried hydrocarbons, such as methane, ethane, and other low-density gases. Deep within the still-soft sediment, the gas molecules are often trapped inside little cages of ice molecules, called clathrates. If you bring these up to the surface, you can set the little ice balls alight by burning the methane inside.

If the hydrocarbons, which are carried in dissolved groundwater, bubble up to the sediment surface at a steady rate, invertebrates and bacterial mats can establish long-term ecosystems, even in areas

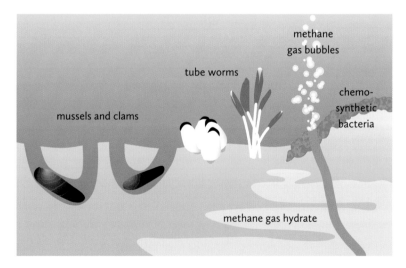

Diagram of a methane seep colonized in a concentric fashion around the seep, with clams under the ocean floor and mussels, tube worms, and chemosynthetic bacteria on the seabed surface. (Redrawn from illustrations at ScienceLearn.org and ScienceInSchool.org)

with very little oxygen. These are somewhat similar to bacteria- and archaea-based communities found at hot deep-sea fluid vents, but the methane seep fluid in the muddy sediment is cold. Photographs taken from ocean submersible vehicles show that seeps are also inhabited by scavenging crabs, shrimps, and snails and that fish move in and out of the low-oxygen areas feeding on detritus and dead animals. Some seeps also have spectacular colonies of giant red-tipped tube worms that are more than 6 feet (2 m) high. Off Vancouver Island are also huge reefs of glass, or silica, sponges, long thought to have gone extinct in the Middle Jurassic.

Cenozoic fossil hydrocarbon seep sites occur in the marine sediment in western Washington, northwestern Oregon, and the outer edges of Vancouver Island. Seeping hydrocarbons changed the chemistry of the very cold seawater so that calcium carbonate precipitated to form limestone, which is extremely unusual because limestone normally forms in warm, shallow tropical seas. In addition, the limestone has another curious feature; instead of forming vast flat beds, it solidified inside large and small tubes and burrows and into mounds within and on top of the sediment, as well as in scattered pebble-size blobs. Limestone is very rarely found in the thick deposits of marine sand and mud of western Washington, so these structures and the assemblages of unusual chemosymbiotic invertebrates led paleontologists to realize that Washington has a wealth of Cenozoic fossiliferous methane seep sites.

In those large ancient seep sites in Washington that are now limestone blocks, the unusual clams are preserved within or close by the limestone, together with glass sponges and, rarely, with tube-worm tubes. Also caught in the sediments were more mobile animals, including snails and crabs. At those sites that had more diffuse fluid flows, the three characteristic fossils are *Acharax dallli*, *Lucinoma hannibali*, and *Conchocele bisecta*, frequently associated with species of the large round gastropods *Liracassis rex* and *Liracassis apta*, species of top shells *Bathybembix*, and the nautiloid *Aturia angustata* (see profile 12).

One question puzzles paleontologists: How did the snails and crabs live in that toxic water? Unlike the clams, which are not very mobile,

The small top shell, *Bathybembix washingtoniana*, is one species of deep-marine top shells found in the marine Oligocene rocks of western Washington. This specimen, 2.5 inches (5 cm) long, was collected on Bainbridge Island by Charles E. Weaver in 1910. The outer shell has been eroded, and the inner pearly or nacreous layer is now a distinct part of the fossil.

perhaps these snails and crabs could move in and out of the region as the fluid flows waxed and waned.

Early fossil collectors studying western Washington marine faunas, including Charles E. Weaver between 1909 and 1950, did not understand the significance of these fossil-rich molluscan assemblages. Not until 1984, when paleontologist Carole Hickman published a landmark paper, did researchers begin to understand these unusual fossils. Her study described six unusual and distinctive invertebrate fossils assemblages that reoccurred in many of the cool- and deep-marine sedimentary rocks from the Eocene-Oligocene in western Oregon and Washington. At the time, paleontologists did not know about methane seeps, but within a few years this new world of a chemical-based ecosystem was being described from living groups of remarkably similar mollusks. A few years later, paleontologists recognized the fossil *Acharax* assemblage, along with fossils of the giant deep-sea clam *Calyptogena* and the deep-ocean mussel *Bathymodiolus* in Eocene to Pliocene rocks in western Washington. All these species are part of the hydrocarbon-based ecosystem, and they now easily identify ancient methane seeps.

A 1966 publication of a compendium of limestone resources of Washington (because limestone is a valuable mineral resource used for cement and for the pulp and paper industry) later led Jim and Gail Goedert to search for seep faunas in Cenozoic rocks of western Washington. They located all the richly fossiliferous large seeps and also found an enormous treasure of marine mammal and bird fossils in these limestones.

More recently, this book's coauthor, Liz Nesbitt, was working with paleontologist Kathleen Campbell, who was one of the first to recognize isolated Pliocene seep sites on the beautiful Washington coast on the land of the Quinault Indian Nation. They were describing the fossils and the geology of these rocks when Liz remembered seeing museum drawers full of these seep-signature clams, *Acharax*, *Lucinoma*, and *Concochele*, which no one had paid much attention to since Weaver had collected them. His first collection, from Bainbridge Island, was in 1909. Other collections across western Washington included seep-indicator taxa from Oligocene rocks on the Olympic Peninsula and in Grays Harbor, Mason, and Pacific Counties.

Weaver and other paleontologists had used species of these clams as index fossils to differentiate short geologic time spans. The assemblages had been collected from many different places in the Eocene and Oligocene marine rocks of western Washington, including in the Lincoln Creek, Blakeley, and Pysht Formations. With the new and better understanding of these complex seep communities, Liz could see that the assemblages were not time indicators but characteristic of methane and hydrocarbon seep ecosystems over long periods of time. "It was the perfect illustration of the importance of museum collections," said Liz. "They may sit unused and unappreciated for a century. But one day science catches up and shows that museums are like libraries of unsolved stories just waiting for someone to make new connections."

HELMET SHELLS
AND WHALE GASES

After a very warm, ice-free planet in the Eocene, the world changed rapidly at the Eocene-Oligocene boundary, around 34 million years ago. For the first time in more than 200 million years, continental ice sheets developed across Antarctica. Using marine microfossils from deep-sea sediment coring studies, paleontologists can track the flow of dense cold water moving north from Antarctica along the bottom of the oceans. Within a couple million years, the water of all the oceans had cooled considerably, which affected the land by completely changing global temperatures and wind patterns. Although no ice formed in the latitude where future Washington State was located, the climate cooled enough to dramatically alter the lives of the marine invertebrates along the coast of Washington and Oregon. The result was a major extinction, when around 85% of the Eocene marine molluscan species were eliminated.

With the loss of these species, a host of new animals moved in to fill the open niches. Most came from the north, already adapted to colder water. Over time they evolved into new species, now found in the Oligocene marine sediments of the Blakeley, Lincoln Creek, and Pysht Formations in western Washington. Some of the most curious are large rotund snails, 4 to 5 inches (10 to 12 cm) in height shaped distinctly like an ancient Roman helmet. The family Cassidae includes all the helmet shell snails (called cassids), including the species from Oligocene rocks of Washington: *Liracassis rex* and its twin *Liracassis apta*. They are thin shelled, indicating that they lived in a cooler-water environment. In some sites, they occur in huge numbers, with most being exact internal molds of the shell where calcite-rich, cement-like mud filled the shell that later dissolved during the fossilization process.

Liracassis fossils were formerly so abundant in places that paleontologists collected them by the bucketful from the fine-grained calcite-rich siltstones north and south of the Olympic Mountains, often at sites associated with fossil whale bones and wood, as well as fossil methane seeps. Despite, or because of, their abundance, the fossils presented paleontologists with several mysteries. Why did these *Liracassis* species accumulate in numbers in a few places but not in others? Why are they frequently associated with fossils of whale bone and sunken wood? Why are they associated with cold methane seeps when modern helmet shells typically occur in warm water and have thick shells?

Like any good detective, paleontologists studying *Liracassis* started to marshal the relevant facts to figure out what they meant to these mysteries. One of the first items of detective work is to piece together the life of the "species of interest." Paleontologists start by establishing when the plant or animal lived. Radiometric dates can be obtained from volcanic ash layers, but that is

This fossil of the large rotund gastropod *Liracassis apta*, 5 inches (13 cm) high, was collected along with the unidentified whale bone encased in a cemented mudstone seen on page 22, in close association with other large specimens of *Liracassis apta*. They belong in the family of helmet shells and, unusual for this family, the fossils occurred in deep cold-water muds.

very expensive, and pristine Cenozoic ash layers are rare in the Pacific Northwest because of active plate tectonic forces. Marine sedimentary rocks are most easily placed into time slots and time sequences by studying the associated microfossils. The most commonly used marine microfossils for establishing time lines are a group of microscopic protists (similar to amoebas) in the order Foraminifera, which live in the ocean's surface layers (planktic) or on the ocean bottom (benthic). Most have complex multichambered shells, called tests, enclosing a single-celled body. The shells are made of crystalline calcite or of tiny

The shells of microscopic single-celled organisms called foraminifera from Puget Sound. The transparent shells and white shells made of calcite are 0.25 inch (0.5 cm) in diameter. The brown shells are made of minute sand grains that are organically glued. Each shell has multiple chambers increasing in size as the amoeba-like organism grew.

sediment particles stuck together with an organic glue; both varieties preserve very well. The name Foraminifera comes from the Latin *foramin* for "hole" or "window," a reference to how each test is full of minute holes that allow the amoeba-like protoplasm to flow out to capture tiny food particles and pull them back in. (The building stone of the great pyramids contains very large foraminifera, which the Greek geographer Strabo thought were petrified lentils from workers' meals.)

A well-known and widely distributed group, Foraminifera evolved in the Cambrian and radiated across all the oceans into at least 40,000 named species, fossil and living. Many have calcite tests so beautiful that microscopists construct minute art pieces out of them. The ubiquity, rapid rate of evolution, and abundance of foraminifera in each geologic time zone has made them critical in deciphering marine environments, paleotemperatures, and time periods. When paleontologists studied the benthic foraminifera from the *Liracassis*-rich Blakeley Formation rocks on Bainbridge Island, they concluded that the fossils had lived in cold water, deeper than 1,000 feet (300 m), during the late Oligocene. (Still-living foraminifera are also important. For example, because foraminifera are so sensitive to environmental conditions, researchers are using them to assess pollution in Puget Sound.)

The next step for the paleontologist detectives was to figure out how their "species of interest" lived. Most modern helmet shell snails live in the tropics in shallow coastal waters. Living helmets are beautiful large gastropods with heavy shells up to 12 inches (30 cm) long that are especially thickened around the aperture—the round or slit-shaped hole where the snail body emerges. (Cassid shells from West Africa were the preferred shell for carving translucent cameo jewelry in eighteenth-century Europe.) Active predators, helmet shells target sea urchins and sea stars for food and rely on their thick shell to help protect them from other predators, such as crabs, fish, and other carnivorous snails. However, there is a group of little-known deep-water modern cassid snails that live between 800 feet (244 m) and 8,000 feet (2,440 m) deep in the ocean, which have thin shells and no extra ornamentation or thickening around the aperture—basically the description of *Liracassis rex*. Further confirmation that paleontologists had a correct assessment of fossil *Liracassis* came from the association between modern helmet shells and other thin-shelled large fossil gastropods, such as *Neptunia* (whelk), *Fulgoraria* (volute), and small clams in the genus *Acila*, all of which primarily inhabit cold, deep seas.

Paleontologists typically study modern relatives to understand the environment and life history of fossils species. This can't be done with the modern deep-water cassids, because they have been dredged from the seabed and very little is known about how they live in the dark, low-oxygen sea bottom. Paleontologists are additionally confounded by the *Liracassis* fossils because no sea urchins are preserved in the rocks, which raises the question as to what these particular snails ate.

The biggest clue to the Oligocene mysteries was finding large numbers of *Liracassis rex* and *Liracassis apta* associated with fossil methane seep sites and with whale bones and large pieces of sunken wood in the deep-water sedimentary layers. Some helmet shell snails were preserved with their aperture directly attached to the bone or the wood. When a whale dies and the body sinks into the mud on the deep seafloor, the carcass creates an entire ecosystem by providing a sudden bonanza of nutrients in a zone that lacks food and oxygen. Taking advantage of oceanographic submersible vehicles, biologists have repeatedly photographed over many years disintegrating whale bodies, or whale falls, teeming with invertebrates and fish and covered in places with dense mats of the bacteria *Beggiatoa*. The species of clams, snails, crabs, tube worms, and sponges are the same as those that live in areas where methane gas emerges from the sediments onto the seafloor. In these unusual environments, these invertebrates survive by utilizing the methane gas and hydrogen sulfide released by bacteria from the decaying body of the whale, which also becomes an attractive location for crowds of hungry scavengers.

Fossil evidence shows that *Liracassis rex* evolved at the same time that early whales were evolving in the northern Pacific Ocean. The oldest whale fossil in Washington, and the rapid radiation of early whales along our coast, coincides with the radiation of large *Liracassis* species that had evolved from a much smaller species in the middle Eocene. Studying the Oligocene *Liracassis* preserved directly in contact with fossil whale bone, paleontologists proposed that these cassid snails were moving along the bone surface munching on the other invertebrates and perhaps even the bacterial mats that were slowly disintegrating the whale carcass. This environment would have had high levels of hydrogen sulfides and very low oxygen, as well as very little, if any, light—a difficult environment for a snail. There must have been sufficient oxygen for them to at least feed, and most likely everything happened slowly.

In addition, the sedimentary layers with the abundant *Liracassis* shells also contain other fossil invertebrates that typically live in low-oxygen, methane-rich environments. Calling on prior studies of sedimentary geology, microfossil paleontology, modern invertebrate biology, and the newly emerging knowledge of how whale bodies decay, Liz Nesbitt and her colleagues arrived at a conclusion: the huge numbers of thin-shelled *Liracassis rex* and *Liracassis apta* congregated around a novel food source in the deep, dark, cold sea off the Washington coast many times in the late Oligocene. Early toothed and baleen whales proliferated along the Washington coast in the newly cooled North Pacific Ocean, probably because there were new food sources. Their subsequent deaths then opened up novel ecosystems in which some newly evolved invertebrates flourished.

Washington State is famous for its whales, fourteen species of which spend part of the year along the coast, in the Strait of Juan de Fuca, or in Puget Sound. Most famous are the killer whales, or orcas. Others include the biggest species to have lived on the planet, blue whales; the species made famous by Herman Melville, sperm whales; the whale order's most famous songster, humpback whales; and one of the world's smallest whales, pygmy sperm whales. All belong to one of two groups of living whales. The species-rich group called odontocetes includes toothed whales, porpoises, and dolphins. The mysticetes includes the sixteen living species of giant baleen whales, such as fin, Sei, minke, and gray whales.

All whales, living and fossil, are in the mammalian order Cetacea, from the Greek word meaning "sea monster." Whales are marine mammals that breathe air but in all other aspects are tied to the ocean. They have elongated bodies, forelimbs developed into fins, no back legs, and a very strong, flexible tail used for swimming. The key features that distinguish whale skulls from all other mammal skulls are the nostrils set far back on the top of the skull, the eyes situated below that, and the thickening of bone around the ear, called the tympanic bulla, which isolates the air inside the skull to improve underwater hearing.

Unlike the odontocetes, baleen whales have no teeth; instead, they feed using plates of baleen that sieve food out of prey-laden water. Despite not having teeth, mysticetes evolved from toothed ancestors, and it was assumed that the earliest species had both teeth and baleen until the massive baleen plates took over the mouth entirely. Two Pacific Northwest fossil whales, however, tell a different story. They are *Sitsqwayk cornishorum* from the Olympic Peninsula and *Maiabalaena nesbittae* from the Newport area, Oregon. Both are Oligocene in age.

Because baleen is made of keratin, like our fingernails and hair, and rarely fossilizes, the transition from the earliest toothed whales to filter feeders with baleen was a mystery. Based on the few fossils that paleontologists did have, they estimated that this major evolutionary breakthrough occurred sometime in the Oligocene. Then field collectors began to find treasure troves of fossil whale skulls in the Pacific Northwest, Japan, and New Zealand. Aided by the addition of so many skulls and so many different features, paleontologists now had many possible but different explanations of how mysticete whales evolved during this critical time period. It has taken years to both prepare the bones from the well-cemented hard rock and for marine-mammal specialists to piece together the complex stories provided by all these new fossils; many more unprepared skulls in museums wait further study, so the story should become clearer.

In terms of geologic time, the evolution of marine mammals is a relatively recent story. The long reign of marine reptiles that lived across all the oceans through the entire Mesozoic (186 million years in duration) ended with the asteroid that also killed the non-avian dinosaurs. This empty marine niche was filled by land-dwelling mammals that modified their bodies to live in and become predators in the seas. It was an enigma to early biologists because they did not have fossils to show the evolutionary steps from land to water. European paleontologists in the early 1800s found fossil whales that looked like primitive whales, but that did not explain how land-living, air-breathing mammals became obligate marine mammals. Not until the 1980s when researchers found a wealth of stunning fossils in Pakistan were they finally able to piece the story together.

Features of the skull and ankle bones illustrated the rapid evolution from small land-living carnivores through a series of steps to a fully marine whale, in a mere 10 million years, in the early Eocene. These primitive whales soon diversified and spread across the oceans. The late Eocene and Oligocene was a bonanza time for whales, and the high diversity of species makes the evolutionary tree look like a tangle of bushy branches, which keep shifting.

Sitsqwayk cornishorum is one of those whales on this complex line leading to baleen whales. In 1993, amateur fossil and mineral collectors found the fossil in the Pysht Formation on the northern Olympic Peninsula coast. These late Oligocene rocks have also produced thousands of marine invertebrate fossils, all well cemented in very hard rock. Paleontologists from the Burke Museum collected the skull, 2 feet (61 cm) long, and, most unusually, almost all the bones of the spine and forelimbs. Bruce Crowley, the museum's fossil preparator, spent the next four years very carefully freeing the softer fossil bone from the cement-hard rocks in an exhibit gallery in the museum because there was not enough room in his usual workspace in the museum's basement. The almost complete fossil is now on prominent display at the Burke Museum.

The primary work on studying this unusual species was done by Carlos Mauricio Peredo and Mark Uhen, who have continued to focus on the amazing whale fossils from Washington. They created the generic name *Sitsqwayk*, writing that it "is [a] Klallam word for a powerful spirit from far out in the water said to bring wealth. Use of the Klallam language is done here with permission from the elders of the tribe." The species name honors the original collectors, John and Gloria Cornish.

Sitsqwayk is large for an Oligocene whale, about 15 feet (4.5 m) long, with strong front flippers. The elongate skull has many skeletal features of mysticete whale bones, including nostrils on the top of the skull and a relatively small tympanic bulla. Neither the upper or lower jaw had teeth, nor was there evidence that *Sitsqwayk* had any baleen. From the arrangement of the skull, Peredo suggested that it was a primitive baleen whale, sitting close to the junction in the evolutionary tree where the mysticetes branched off from the toothed ancestors. What is known now is that the first baleen-bearing whales originated in the Oligocene, most likely from the oldest toothless mysticetes, which are *Maiabalaena* and *Sitsqwayk*.

In 2018, Peredo and colleagues published a paper on another fossil that also had no teeth and no evidence of baleen. Nearly 10 million years older than *Sitsqwayk*, it is from rocks exposed near Newport, Oregon. The fossil was collected in the 1970s by Doug Emlong, a legendary Oregon fossil hunter who, as a teenager and young man in the 1960s and 1970s, scoured the beaches and eroded cliffs along the Oregon coast and occasionally into Washington. At least fifteen of

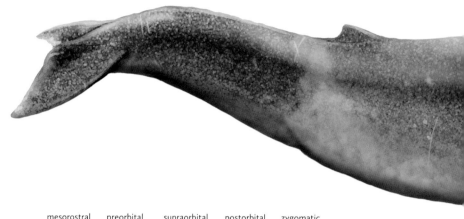

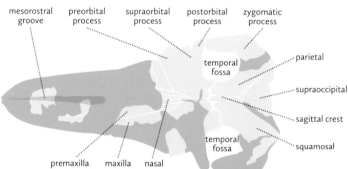

mesorostral groove · preorbital process · supraorbital process · postorbital process · zygomatic process · temporal fossa · parietal · supraoccipital · sagittal crest · squamosal · temporal fossa · premaxilla · maxilla · nasal

Skull of *Sitsqwayk cornishorum*, which had no teeth nor evidence of baleen plates in its jaw. The skull and lower jaw are original fossil material; the long upper jaw was reconstructed to fit for an exhibit at the Burke. This is one of the very few early whales that was found with the entire skeleton as well as the skull.

Reconstruction of the oldest mysticete whale, *Sitsqwayk cornishorum*. Illustration by Gabriel Ugueto commissioned for the Burke Museum, used with permission.

his marine-mammal finds were new genera or even new orders; subsequent paleontologists have used Emlong's fossils to reconstruct the evolutionary lineages of whales, seals, and extinct marine mammals.

Emlong sent his specimen to the Smithsonian Institution, where it sat on a shelf encased in hard rock until paleontologists realized its significance. Peredo and Nicholas Pyenson decided to use state-of-the-art computerized tomography (CT) scanning technology to image the fossil within the rock and discovered the significant features of a narrow jaw with no teeth and no attachment sites for baleen shafts. They named the new genus the "mother-whale" *Maiabalaena* and the species *nesbittae*, which honors this book's lead author, Liz Nesbitt. This fossil is sister to *Sitsqwayk* in the whale evolutionary tree, and the two species indicate that mysticete whales lost their functional teeth entirely before evolving baleen.

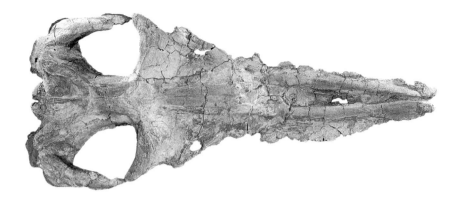

Dorsal view of the almost complete skull of *Maiabalaena nesbittae*. The blowhole (nostril) is situated on the middle of the snout. Large muscles attaching the lower jaw to the skull occupied the holes in the back of the skull. The eye sockets were situated in approximately the middle of the skull, low down and not visible from this angle.

The obvious question is, How did such large animals get their food? Baleen whales have very strong jaw muscles and retractable tongues and can open their giant mouths to gulp in huge volumes of seawater plus their prey—fish or krill. Marine-mammal paleontologists studying the physics of modern baleen whales concluded that early toothless mysticetes ate by rapidly opening their mouth underwater and using their tongue and throat muscles to generate suction, thereby pulling smaller prey into their mouth, a process known as suction feeding.

Three other fossil whales from Washington play important roles in this story. *Fucaia buelli*, from the Makah Formation near Neah Bay, is about 30 million years old. The generic name refers to the Strait of Juan de Fuca. *Salishicetus meadi* was collected in 1973 from the Lincoln Creek Formation in Rochester, Thurston County, and is around 25 million years old. This partial skull includes the lower jaw and the tympanic bulla. These two fossil whales are much smaller than *Maiabalaena* and *Sitsqwayk* and, unlike the larger whales, have elaborate teeth with sharp edges and many pointed cusps. A third skull, *Borealodon osedax* from the Pysht Formation, falls in a completely new group of ancestral toothed mysticetes. This skull was first noted because of traces of holes made by bone-eating worms, evidence that it was part of a whale that sank to the bottom of the deep, very low-oxygen (anoxic) seas. Evidence suggests that these dolphin-sized whales, 6 to 10 feet (2 to 3 m) long, were fierce predators that caught fish, then used their specialized teeth to cut and chew in the same manner as land carnivores such as dogs. *Fucaia* and *Salishicetus* are contemporaneous with *Maiabalaena* and *Sitsqwayk*, respectively, telling us that toothed mysticetes lived alongside toothless mysticetes for much of the Oligocene before eventually going extinct around 23 million years ago.

In the past decade, Peredo, Uhen, and Pyenson and their students have named seven new species of whales from the Pacific Northwest. Because of their work, Washington has emerged as a center of research in the modern understanding of whale evolution, particularly in regard to early mysticetes. It is arguably some of the most exciting, promising, and important paleontological work taking place in the state.

Whales have perplexed humans for thousands of years. Aristotle called them the "strangest of animals," and up until the nineteenth century, people still debated whether whales should be classified as mammals or fish. More recently, the questions have centered on whale classification and evolution. In particular, researchers have long tried to better understand how and when whales separated into their two suborders, Odontoceti, or toothed whales, and Mysticeti, or baleen whales. Fortunately, some of the more intriguing evidence for paleontologists has been found recently in Washington.

Modern toothed whales include orcas, dolphins, porpoises, beaked whales, and sperm whales. Active predators, they have a streamlined body with a flattened tail designed for speedy swimming. Their adaptations to a permanent life in water have altered the skeleton to such a degree that it no longer seems to be a mammal. The shape of the bones in the whale skull are greatly rearranged from their early ancestors' land-living skull shape. On the top of the skull, the bones have telescoped for an air-breathing life in water. The jaws have elongated, the nostrils have moved back and onto the top of the skull, and the relatively small eyes lie below the blowhole and close to the jaw margin. Surrounding the large brain are well-knit thick bones, supplemented by neck vertebrae fused to give greater stability. In front of the brain—what we would call the forehead—the skull bones close to the blowhole form a concave dish, which holds a large fatty mass called the melon. This is part of the sophisticated echolocation system in all odontocete whales.

Another key characteristic that separates odontocetes from baleen whales (such as humpback and right whales) is their ability to echolocate. Mysticetes, in contrast, lost the ability to echolocate as they developed the novel architecture of baleen plates in the jaw. In order to detect prey, other whales, and the shape of features of the seabed, odontocetes generate a range of powerful clicking sounds within an air sac in their skull, at much higher frequencies than the clicks and

TOP Skull of the modern spinner dolphin, *Stenella longirostris*, 30 inches (76 cm) long, showing the very long snout with a row of small pointed teeth. The skull is domed to accommodate the dolphin's very large brain and the melon—fatty tissue used in echolocation. The blowhole made from the combined nostrils is far back on the skull. The eyes are situated low on the side, in line with the teeth.

BOTTOM Fossil long-nosed dolphin skull, 35 inches (88 cm) long, from Clallam County, is still unnamed. The snout was reconstructed from numerous broken bone pieces in the rock. The underside of the jaw shows many small sockets for simple teeth, but the last third of its length had no teeth. The skull shows distinct evidence that early dolphins used echolocation.

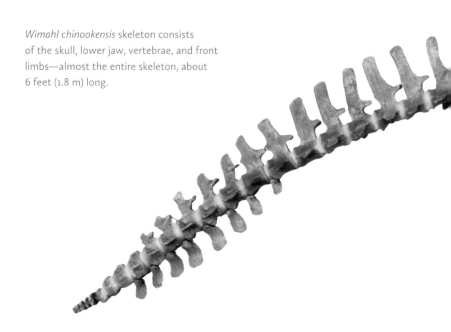

Wimahl chinookensis skeleton consists of the skull, lower jaw, vertebrae, and front limbs—almost the entire skeleton, about 6 feet (1.8 m) long.

songs recorded from their underwater conversations. Whales enhance and focus these sonic waves by passing them through their fatty melon. The clicks then bounce off the target and are picked up by elaborate ear bones, which, unlike ours, are isolated in air-filled cavities. Fatty tissue on the lower jawbones further focuses the returning sound waves toward the ear bones—acting like the outer fleshy ear of land mammals.

Odontocetes can see, but echolocation is much more reliable in the water and much better for murky or very deep water. This remarkable way of navigating the underwater world began in the earliest ancestors of whales when they first became aquatic, 48 million years ago.

Odontocete teeth are also unusual, not just in comparison to the plates of baleen whales but also in the mammalian world. Other mammals, such as humans, have different-shaped (heterodont) teeth in each jaw: incisors, canines, and molars that fit neatly together top to bottom. They evolved for cutting in the front and chewing in the back. In contrast, odontocete teeth are all sharply pointed simple cones with gaps

Reconstruction of the small long-nosed dolphin, *Wimahl chinookensis*. Illustration by Gabriel Ugueto commissioned for the Burke Museum, used with permission.

between, which facilitates catching fish and squid that they swallow whole. Even orcas that eat seals do not chew their catch.

Many toothed whale fossils have been found in marine rocks of western Washington. Most of these are partial skulls and most have not been described. However, there is one outstanding example that has been named and described. *Wimahl chinookensis* was collected by Jim and Gail Goedert from an island near the mouth of the Columbia River. Unlike the vast majority of fossil whale finds, *Wimahl* consists of skull, lower jaw, vertebrae, and front limbs—almost the entire skeleton. Because the Columbia River is tidal at its mouth, Jim had to choose a time with an especially low tide, get across to the island, dig out the fossil, quickly wrap the many pieces, and bring it all back before being flooded by the next tide. They had only six hours.

Wimahl chinookensis is a small long-nosed dolphin, about 6 feet (1.8 m) in length, that dates from the early Miocene, around 20 million years ago. The long, narrow top and bottom jaws are full of small pointed teeth that are the same size and look similar. Those on the tip of the bottle-nosed jaw interlock top to bottom like a vise. The back ones do not. The ear bones show that this animal could echolocate. *Wimahl* belongs in the fossil dolphin family called Kentriodontidae, with fifteen other genera, and is the oldest and northernmost member of the family found in the Pacific Ocean. Its genus name reflects the Chinook peoples' name for the Columbia River and together with the species name roughly translates as the "big river of the Chinook."

One of the oldest odontocetes is *Olympicetus avitus* from the northern Olympic Peninsula, around 30 million years old. This new fossil genus and species includes two juvenile skulls collected in 1985 by Jim and Gail Goedert. Unlike modern dolphins and *Wimahl chinookensi*, this small dolphin had a shorter skull and heterodont teeth. The teeth in *Olympicetus* are longer than those of modern dolphin teeth, with a series in the back of the jaw somewhat similar to typical mammalian molars. The other teeth have a row of little tubercles across the top that look like a crown. *Olympicetus* also shows evidence that the skull was used for some sort of echolocation.

Coastal rocks in Washington, Oregon, and Alaska have yet another group of curious odontocete fossils—extinct dolphins related to modern river dolphin. Only two species of freshwater dolphins in the superfamily Platanistoidea still exist: the endangered Ganges River dolphin and the Amazon River dolphin. (A third species, Yangtze River dolphin, was declared extinct in 2007.) In contrast, during the latest Oligocene and the Miocene, this superfamily was very diverse and widespread across the Pacific Ocean. The platanistoid group includes the small long-nosed dolphins from Washington, *Squaloziphius emlongi* (honoring Doug Emlong) from the early Miocene Clallam Formation, and *Yaquinacetus meadi* (honoring the Yaquina people, the Indigenous people who lived where the fossil was found) and *Perditicetus yaconensis* (named for the Yacona people, native to the central Oregon coast) from the Nye Formation, south of Newport, Oregon. Also, from the middle Oligocene Poul Creek Formation, in Yakutat City, Southeast Alaska, is *Arktocara yakataga*, the northernmost and oldest platanistoid known. The latter three species reflect a new awareness within the scientific community of the importance of recognizing and acknowledging the Native people who long inhabited and continue to inhabit the land where fossils were found.

Many of these Miocene platanistoid dolphins have exceptionally long snouts (elongated jaws), some more than twice as long as those of living river dolphins and resembling the "swords" of swordfish and billfish. Some of the fossil dolphins have a lower jaw shorter than the upper jaw and no teeth on the end of the upper jaw. To find out how these animals used their very long snouts, paleontologists borrowed applied-beam theory from structural engineers to estimate the load-bearing capabilities of the measured cross sections of the beam-like bony jaws. Whales evolved very short, compact neck vertebrae that prevent the head from moving independently from the body—a hydrodynamic innovation. However, these fossil dolphins with long snouts evolved with neck vertebrae that are relatively longer than most and have articulation showing that they had a mobile neck, like land vertebrates do. Researchers concluded that the dolphins slashed at fish with their long snout to stun them while speeding through schools in the shallow water, then circled back to pick up the prey and swallow them whole.

One of the paleontologists studying and describing dolphin fossils from the Pacific Northwest said that the most exciting aspect of studying these fossils is their staggering diversity around the world during the Miocene. Washington's cetacean fossil record is especially rich from Oligocene and Miocene marine rocks, and there are many more skulls collected and awaiting preparation and names, descriptions, and placement on the evolutionary tree. It bodes for an exciting future for whale paleontologists.

Paleontologists have many ways to reconstruct the lives of the animals they study. Until the past few decades, they relied on observable physical evidence, such as comparing fossil anatomy with living relatives, looking at the associated sediments to understand the paleoenvironment, or examining tracks, burrows, and other traces of the animal's existence. More recently, paleontologists have begun to employ a high-tech arsenal using ancient DNA, sophisticated computer modeling, and X-ray analysis to flesh out the lives of extinct animals. One of the better illustrations of these changes in paleontology comes from one of the more unusual fossils in the state: *Kolponomos clallamensis*. The *Kolponomos* fossils also illustrate the importance of nonacademic collectors to the science of paleontology.

In 1957, Bette Willison—a schoolteacher who found numerous fossils subsequently studied by researchers, including the unusual river dolphin *Squaloziphius emlongi* (see profile 16)—unearthed a partial skull with the roots of several teeth near Slip Point, in Clallam County, about 20 miles (32 km) east of Neah Bay. She donated it to the Burke Museum along with associated clams and snails from the same rocks; these fossils showed that the skull was earliest Miocene in age. Subsequent work on the unusual fossil by paleontologists at the University of California, Berkeley, led them to conclude that it was a carnivore. The animal appeared most similar to a small bear, with a small—5 inches (13 cm) long—skull, but unlike other carnivore fossils this one has a distinctly downturned snout and large flat teeth akin to the teeth of otters. The Berkeley researchers named the species *Kolponomos clallamensis*.

A dozen years later, additional similar-looking skull bones were found on the Oregon coast, near Newport. The collector was Doug Emlong. His 1969 discovery included *Kolponomos* skull parts, dissociated vertebrae, ribs, and foot bones. Then in 1976, he found a concretion on the same beach and remarkably realized that it was a broken piece from the original one he had found containing *Kolponomos* remains. Emlong sent his marine-mammal fossils to the Smithsonian Institution, where researchers named the new fossil *Kolponomos newportensis*.

Skull of *Kolponomos clallamensis*, 5 inches (13 cm) long, with molars suitable for crushing shells. Overall, the skull resembles that of living bears and otters: wide at the back to hold big jaw muscles, narrow across the nose. The snout (left side of image) is downturned compared with that of otters and has short pointed incisor and canine teeth.

Along with Bette Willison's Slip Point fossils, they are the only fossils known from this genus. These two species have no other relatives and left no descendants, at least none that have been found.

Kolponomos had a robust, bearlike skull with spaces for very large jaw muscles and eyes directed forward rather than on the side of the head as in living bears. The end of the snout was downturned, with front teeth set into thickened bone at the end and a particularly deep lower jaw, like a chin. The jaws were set with otter-like flattened low molars. Researchers concluded that these features allowed the animal to pry clams off the rocks in the intertidal zone. As occurs with otters, which hammer and crush clams, snails, and sea urchins with their flat molars, *Kolponomos* smashed the shells in order to obtain the edible meat within. Now that the researchers had fossils bones from the feet, they could further tell that *Kolponomos* walked like a bear and most likely lived along beach areas, prying invertebrates such as mussels and barnacles off the rocks and eating them. Thus, *Kolponomos* were land-living animals that fed off the sea. The latter conclusion comes in part from the bones' location in marine sediment rich in mussels, giant scallops, and a wide variety of clams and snails. In the Miocene, the Pacific Northwest beaches provided a smorgasbord of perfectly suitable food for them.

Reconstruction of a *Kolponomos* eating a clam. Based on the skull length, the animal must have been about the same size as a large sea otter, but only the skull of *Kolponomos* is known.

Now jump to 2016, when the next published paper on *Kolponomos* was written by Jack Tseng and other mammalian paleontologists at the American Museum of Natural History. Their goal was to investigate the animal's unusual feeding style of biting, then pulling, mussels off the rocks, or what Tseng describes as an "unsolved mystery both taxonomically . . . [and] morphologically." In particular, the paleontologists wanted to better understand precisely how the carnivores removed their proposed food source, the mussels that were attached to the beach rocks.

Tseng and his colleagues were able to undertake this study because of recent advances in technology. One was X-ray computerized tomography (CT scans). Utilizing techniques employed by medical technicians, paleontologists can now "see" inside the rocks and create incredibly detailed digital images of the fossils inside. Once a 3-D image is obtained, the fossil can be 3-D printed and a replica reproduced for further studies or can be studied on a computer screen, which is what Tseng's team did. Using these true-to-life models allowed them to test the strength and flexibility of the skulls and to compare the bite motion and pressure of saber-toothed cats (*Smilodon*) and other extinct carnivores, which led them to conclude that *Kolponomos* had a bite like the famous extinct cat.

The researchers propose that the long-extinct beach dwellers used their large neck muscles and strong thick lower jaws to anchor and wedge their teeth under their prey. They then closed their mouth around the shell and torqued their skull to dislodge their meal, using their "chin" as a fulcrum. (Anyone who has tried to harvest mussels at low tide knows how difficult it is to pry them lose. Materials engineers have begun to try and develop better adhesives based on the complex combination of proteins that mussels use to adhere to underwater surfaces.) The computerized models further showed that *Kolponomos* crushed their prey, though in a different style than an otter. They also had a different eating method than other large mammals that eat mollusks. Walruses primarily dig them out of the sand and suck the body out, leaving the shell behind, whereas otters break the clamshells with stones or crush smaller ones with their molars. No modern animals employ the *Kolponomos*'s unusual feeding habits.

Although *Kolponomos*, otters, and saber-toothed cats are all within the order Carnivora, they evolved within different families. "We [carnivoran systematists] still have a lot of trouble placing *Kolponomos* on the carnivoran tree," says Tseng. "Its dental and cranial features form unique trait combinations, which do not help much with comparisons to other possible relatives or ancestors. The best we can say is that *Kolponomos* is likely related to bears and bearlike carnivorans. Its ancestors probably were not specialized hard-food crushers, but we still don't know when this evolution toward the feeding behavior we hypothesized occurred."

Kolponomos were not alone as very odd-looking beach-dwelling marine mammals on the Oligocene and Miocene beaches of the Pacific Northwest. Not commonly found here, but widespread from California to Japan, the other unusual seaside inhabitants belong in their own order, Desmostyolidae. At present, paleontologists have yet to resolve their evolutionary relationship to other herbivorous mammals. Desmostylids have skull and jaw shapes that place them close to manatees, as well as to elephants or possibly the group that includes rhinoceroses and horses (perissodactyls). None of these curious mammals, however, exist at present; all desmostylids went extinct after a short geologic history.

One of the best known is *Desmostylus hesperus*, which lived in shallow coastal waters and estuaries. The adults weighed about 440 pounds (200 kg), were 5 to 6 feet (about 2 m) long, and looked like a small hip-

A desmostylid tooth, with a strong pillar-like structure unlike any other mammal tooth known, 4 inches (10 cm) high. Desmostylids were large herbivorous mammals that may have lived like hippopotamuses in lakes, estuaries, and even on the seashore, eating aquatic plants.

popotamus with sturdy short legs but a more elongate narrow skull. Desmostylids also had lightweight bones, which has led paleontologists to assume that they lived in the water and swam by dog-paddling with their front legs like polar bears do. Their most distinctive feature were their molars, which gave the order its name: *Desmostylus* translates as "bonded pillar," in reference to the fused round pillars that make up each molar. These molars are so unusual that even a broken bit is instantly recognizable to paleontologists. In addition, the molars erupt in sequence; as with elephants, when the first tooth is worn down, the next one erupts. With their pillar teeth, the herbivorous *Desmostylus* ate sea grasses, though their unusual front teeth (incisors and canines), which extended as short tusks, may give the impression they ate more formidable food. Geochemical studies of the teeth show that they lived in rivers, estuaries, and shallow seas.

In Washington, paleontologists have found only Miocene-age teeth, whereas in California entire *Desmostylus* skulls have been recovered. However, three older desmostylid genera have been described from late Oligocene rocks, along the Strait of Juan de Fuca and on the beaches of central Oregon. *Cornwallius sookensis* was found on the southern end of Vancouver Island, along with the remains of *Behemotops proteus*. Both were first collected by Doug Emlong, in the Pysht Formation. The name *Behemotops* is derived from *b'hemah*, which means "beast"; some etymologists suggest that the word comes from an Egyptian word for "hippopotamus." The paleontologists who bestowed the name were equally perplexed by its origin but perhaps ultimately agreed with the medieval poet John Lydgate who wrote of the behemoth that the name "doth in Latin playne expresses a beast rude fell of cursedenesse."

More than five hundred species of birds have been reported from Washington State, ranging from exquisite hummingbirds to majestic eagles. Collectively they total billions of individual birds that have graced our skies. Few of them, though, will ever fossilize, partly because bird bones, which are generally hollow to facilitate flight, are so fragile. That has certainly been the case with birds from the past; their fossils are very rare. To our good fortune, some unusual and fascinating bird fossils have been found here—in particular, in the Oligocene marine rocks of the northern Olympic Peninsula. The bones survived, in part, because they were less fragile and more robust and came from birds less focused on the air and more on the water.

The most impressive bones are from an extinct group that paleontologists characterize as giant flightless penguin-like birds known as plotopterids. Reconstructions of the animals show that some were larger than penguins, possibly twice the height of emperor penguins, which stand about 39 inches (1 m) tall and weigh around 65 pounds (30 kg). Like penguins, these giant birds lived on land, either on beaches or sea cliffs. With wings highly modified to form rigid paddles, no flexibility in the wrists, and greatly expanded but thin-boned shoulder-girdle bones adapted to the attachment of very well-developed muscles, plotopterids evolved for life along the coast, where they hunted for fish in the sea. They also had thick-walled, dense leg bones, which provided strength and heft to make them less buoyant.

Plotopterids also benefited from their penguin-like skull with its long pointed beak and slit-like nostrils. Such a body shape maximized the birds' ability to torpedo through the water hunting fish with their wings providing propulsion and steering, the feet adding little to the forward or directional processes. This body shape has enabled living auks (a common Washington coast bird) to be able to dive to 328 feet (100 m) in depth; emperor penguins regularly dive to 656 feet (200 m) and have been recorded at more than 1,640 feet (500 m) down. Like penguins, plotopterids stood upright and probably were not very agile on land.

Reconstruction of numerous plotopterid birds on the beach. These large flightless birds had an upright gait and swimming style similar to that of penguins, but they are more closely related to cormorants and darters. The fossils have been found only in the North Pacific. Artwork by Mark Witton.

Amazingly, the initial concept of plotopterids began with a single broken shoulder-bone fossil identified by pioneer avian paleontologist Hildegarde Howard. Nearly four decades after Howard earned her PhD in 1928, she was working at the Los Angeles Natural History Museum when a collector brought her a bone found at a sand quarry in the San Joaquin Valley of California. Making a remarkable insight, Howard recognized that it was a new type of diving bird—combining aspects of two different families, those of penguins and of auks—that used their wings for underwater propulsion. In contrast, other well-known birds that hunt for fish underwater, such as loons and cormorants, use their feet for propulsion. She named the new animal *Plotopterum joaquinensis*, creating the new family Plotopteridae from the Latin *plot*, meaning "swimming," and *pterum*, meaning "wing."

Rare in birds, wing-propelled diving is the result of parallel evolution, which developed within different lineages for the same type of lifestyle. Paleontologists studying ancient birds have shown that plotopterids are closer in evolutionary lineage to the families that contain cormorants, gannets, and darters than to penguins. The theory for wing-propelled diving for both plotopterids and penguins is that they each evolved separately from flying ancestors that had developed a feeding strategy of plunging into the sea. Plotopterids appear to be limited to the North Pacific rim, whereas living and fossil penguins are confined to the southern hemisphere. The oldest plotopterid described is from the late Eocene Keasey Formation of northwestern Oregon, and the youngest is from the early Miocene from central California. Some paleontologists also speculate that plotopterids became extinct with a rise in numbers and diversity of seals and dolphins competing for the same food.

In 1977, Doug Emlong collected the partial skeleton of Washington's first plotopterid and sent it to the Smithsonian Institution. It was named *Tonsala hildegardae* (*tonsala* means "oar wing"). Since that initial work, other plotopterid fossil bones have been collected from late Eocene to early Miocene marine rocks in Japan and all along the West Coast of North America. All the finds are of partial skeletons, or even single bones, encased in very hard rock. These include many additional *Tonsala* specimens, as well as five other genera from western Washington, northern Oregon, and Vancouver Island, British Columbia. Based on these discoveries, paleontologists refined the description of plotopterids to its modern description.

In 2017, paleontologists published a paper describing bones of another large marine-bird fossil from the banks of the Columbia River, near Knappton, Washington, in Pacific County. The limb bones, pelvis, and some vertebrae are all similar to those of modern albatross (genus *Diomedea*), but much smaller. Named *Diomedavus knapptonensis*, they are the oldest albatross fossil bones known in the North Pacific, at 20 million to 24 million years old.

Many of the fossils used to describe plotopterids were collected by Jim and Gail Goedert. Jim found his first plotopterid fossils in the Pacific Northwest, *Phocavis maritimus*, in Columbia County, Oregon. Jim and

Bones of *Tonsala hildegardae* from marine Oligocene rocks on the Olympic Peninsula. They were carefully prepared from the cemented concretion. Pelvis and fused lower back vertebrae in this specimen (bottom image) is 5.5 inches (14 cm) long.

colleagues have also found and described additional bones from *Tonsala hildegardae* and *Tonsala buchanani*, as well as naming and detailing the anatomy of three new plotopterid species from southwestern Washington: *Olympidytes thieli*, *Klallamornis abyssa*, and *Klallamornis clarki* (the largest plotopterid from North America). These species were named after talented amateur collectors who found the bones and donated them to museums: William Buchanan, Bruce Thiel, and Robert Clark. Another species, *Stemec suntokum*, was found near Sooke, on southern

Vancouver Island in the same rocks that produced the desmostylid *Cornwallius sookensis* and the primitive toothed baleen whale *Chonecetus sookensis*. This plotopterid species was also named after amateur collectors, Leah and Graham Suntok, who donated the fossils to the Royal British Columbia Museum in Victoria.

Although Jim Goedert's name appears on most of the scientific papers, his work has been collaborative with his wife, Gail. Together, they spent many years and untold miles walking along beaches and riverbeds looking for fossils in western Washington and Oregon. The information obtained from their collections of thousands of marine vertebrate and invertebrate fossils, along with the detailed records they kept on locations, features in many of the stories in this book. Never planning to keep their fossils in a garage, they always wanted them to be studied and donated to museums. Jim initially sought out the relevant specialists who would be interested in the Goederts' discoveries and later became one of the experts describing and naming new fossils. He has worked with many professional paleontologists from across the United States, Europe, New Zealand, and Taiwan, all while holding a job in the private sector. His most celebrated finds include the plotopterid specimens, as well as a suite of early whale fossils and unusual invertebrate faunas from methane seep sites. His name is on well over a hundred papers published in paleontological journals, and he is now well known.

One of Jim's many talents is the way he shares his expertise and enthusiasm, giving support to other amateur collectors and local museum personnel in Washington. In 2019, after Jim and Gail won the Paleontological Society's Strimple Award, he said, "Getting fossils published has always been our goal. So many amateur collectors say, 'Why should I donate my fossil when it will just be in a drawer somewhere and likely never be displayed?' We reply with, 'Why worry about a particular fossil ever being publicly displayed?' If the specimen gets published and figured (and even better if a 3-D scan is made available) in a scientific journal, the fossil is 'displayed' to anyone around the world with a computer, smartphone, or tablet—potentially millions of people—basically anytime, anywhere, forever."

THE TERROR
OF THE DOCKS

During the late nineteenth century's economic growth in Puget Sound, when Euro-American settlers started to build piers, trestles, and wharves for shipping and rail, one of the biggest problems was a clam known as the "terror of the dock builder." Also called teredo clams, and more colloquially shipworms, they were voracious destroyers of wood, not only in the Sound but around the world. The name shipworm comes from the long—up to 23 inches (60 cm)—body of the clam, which looks like a worm. Using their helmet-shaped shells to gnaw into wood, the teredo clams create elongate holes, or burrows, which can convert a solid piling into a useless structure with a Swiss cheese–like texture.

Around 30 million years ago in the Oligocene, teredo clams also bored their way through wood on the Washington coast. The evidence comes from abundant wood fossils in the deeper-water sedimentary layers that crop out along the Olympic Peninsula and in Grays Harbor and Pacific Counties. In contrast, little wood occurs in the slightly older Eocene marine rocks nearby. The interval between paucity and plenty was a time when two geologic events happened simultaneously along the Pacific Northwest coast. First, the climate changed, after a long period of extreme warmth across the entire world, to very cold. Second, the rise of the ancient Cascade volcanoes affected erosion rates.

One hypothesis is that the new ecosystems favored trees with more-resilient wood, and the steep slopes of volcanoes so close to the shoreline could have created logjams along the Oligocene beaches. Logs floating in the water would quickly attract the floating larvae of wood-boring clams looking for homes. The logs then sank into the mud on the seafloor and fossilized. Very occasionally, the tiny clamshell is also preserved within its very long burrow.

The oldest recorded traces of shipworms in wood are from the Jurassic (about 190 million years ago) of Europe. Two groups of boring clams still leave their marks, one in wood and one in wood and rock. The teredinids make elongate boreholes into wood and digest the cellulose, whereas the pholads, or piddocks, bore into rock as well as wood and filter food out of the water from the safety of their rocky home.

Teredo clams are unlike any other clam in how they look and live. The living *Teredo* has a very elongate body with two tiny shells and a reduced foot at the base (which is really the front end of the clam). Soon after the larva hatches and settles, it begins to bore into the wood. As it grows, it expands its living space, generally paralleling the direction of neighboring teredo clams. The outer edges of the calcite shells do the actual digging and enlarge the hole as the clam grows. Like other clams, it has two short siphons at the end of the body, through which the clam ingests and egests seawater. Inside the clam's gill tissue are

A block of fossil wood 8 inches (20 cm) wide from the Pysht Formation, Clallam County, riddled with teredo clam (shipworm) borings going in all directions and sometimes overlapping. The log must have been floating while the clams were still alive. The log then fell into the mud and the burrow spaces were later filled with calcite and quartz during the fossilization process.

A 50-million-year-old Eocene leaf fossil from Stonerose fossil sites near Republic exhibits evidence of insect damage. Studies show that insect damage to leaves was much more severe in past warmer global climates.

symbiotic bacteria that produce an enzyme for their host to help digest the cellulose. To prevent their burrows from collapsing, *Teredo* clams line the tubes with a thin layer of calcium carbonate.

Teredo clams have a long history of troubling European waters. The great Roman poet Ovid wrote of his mind wasting away "like a ship infected with the hidden wood-worm." Although biologists consider teredo clams to be tolerant of a wide range of water conditions, when European embayments and estuaries became polluted from industrial waste, the shipworms largely disappeared. Now that these waterways have been cleaned, the *Teredo* are back and happily invading boats and docks again.

Over the past few decades, many paleontologists have devoted their careers to the science of ichnology, from the Greek word for "tracks," because trace fossils provide key ecological information. For example, different types of invertebrate traces are usually confined to different water depths—as were their trace makers. Sedimentary geologists can use the assemblages of traces from the same setting to determine the conditions where and when the sediments were deposited. These trace fossils are very useful when you have hundreds of feet of sedimentary strata and no body fossils.

In addition, fossil leaves, stems, and fruit frequently show insect damage—including chomped leaf edges, rolled leaf buds, and pierced and sucked leaf veins—on the plants, and identifying these traces can increase the list of known insects in a fossil ecosystem. Other trace evidence includes boreholes, gall constructions, and seed predation. Studies show that as climate warmed, as in the Cretaceous and Eocene, insect damage to leaves and wood becomes more prolific than in the later Cenozoic.

An international convention dictates the naming of trace fossils, with specific ichnogenus and ichnospecies names. For example, the teredo clam–generated burrows are called *Teredolites*. Although ichnospecies do not meet the biological standard of true interbreeding species, the names do provide a way to ensure that paleontologists are talking about the same thing.

In marine sediments there are thousands of named traces, most made by invertebrates and not directly associated with any one organism. Burrows can be horizontal, vertical, branched in all directions, in muds, in limestone, on hard surfaces, or in wood. Traces can be found in both freshwater and marine deposits, but they are much more common in shallow marine sandstones and siltstones. They are evidence for a much larger diversity of animals in each assemblage than found when looking only at body fossils. For example, moon snails that bore into clamshells or other moon snails to eat the soft body inside leave behind a telltale signature hole.

Hermit crabs that live in abandoned snail

A juvenile Eocene moon snail from the Cowlitz Formation, Lewis County, measures 1.5 inches (3 cm) high. Moon snails are carnivorous, and the incomplete hole bored into this shell was made by another moon snail. In addition, the aperture of the shell, where the snail's body was, had been crushed and broken by a predatory crab.

shells produce a distinct mark on the bottom of the shell as they drag it across the sand flats. Another easily recognizable trace fossil is found in clam shells and snail shells riddled with tiny holes. These are the attachment holes of sponges that used shells lying on top of the sand as the only available hard substrate for anchorage. The sponges are very rarely preserved, but ichnologists often find their distinct traces, called *Entobia*.

If we are lucky, an animal is preserved within its burrow, which can facilitate a comparison between fossil fauna and their modern counterparts. For example, in the Cenozoic sediments of western Washington, ghost shrimps died and fossilized in their burrows. Today, in Puget Sound and along the Washington coast, ghost shrimps in the same genera as their long-extinct relatives live within densely packed burrows, like layers of apartments, which they dig in intertidal sands and muds, mostly near river mouths. Based on the habits of modern ghost shrimp, paleontologists have a better understanding of the rocks with the fossil ghost burrows even without the shrimps that lived in them.

Perhaps the most infamous group of trace fossils are coprolites, or fossil scat. As evidence of what an animal ate, they provide critical information to the animals' biology. One giant coprolite found in Late Cretaceous sedimentary rocks in Saskatchewan, Canada, was so large—18 inches (46 cm) long—that the collectors reasoned the only animal of the right size at that time was *Tyrannosaurus rex*. Not surprisingly, the fossil poop contained crushed bones. Some Cenozoic fossiliferous rocks from Montana, the Dakotas, and Wyoming have fossil coprolites attributed to both herbivorous and carnivorous mammals, as well as coprolites from large fish.

More recent coprolite finds in caves, left over from the last Ice Age, have given paleontologists greater insight into what mammoths, sloths, large carnivores, camels, bison, and even ancient humans ate. Snippets of mitochondrial DNA have been extracted from some of these, and labs have been able to sequence small pieces of the ancient animals' genetic code. Coprolites found in the Paisley Caves in Oregon contain human mitochondrial DNA, and radiocarbon dates place them at 12,750 to 14,290 years before present.

Three coprolites, fossil feces, that were collected in the Oligocene Brule Formation in South Dakota, each 2 inches (5 cm) long. These have tiny bits of bones in them, indicating that they were from a carnivorous animal. Coprolites are not common and often difficult to authenticate.

There is one well-publicized assemblage of so-called trace fossils from southwestern Washington that has made its way into many museum collections and generated numerous contradicting scientific papers. Many look remarkably like vertebrate scat but are actually siderite nodules from the 6-million-year-old Wilkes Formation near Toledo, Lewis County. Siderite is a mineral composed of iron carbonate and is usually yellow-brown to black in color.

Because these nodules so closely resemble scat, hundreds of these "coprolites" have been accessioned into museum collections, placed on fossil collectors' coffee tables, and put on sale in geologic supply houses and on eBay. But no vertebrate bones, which could indicate an animal origin, have been recovered from the Wilkes Formation. Nor is there any organic material in the nodules, although there are plant fossils representing a clay-rich marshy environment of deposition. Plus there are many thousands of other nodules that bear no resemblance to vertebrate feces.

Geologists have hypothesized several ideas for the formation of the nodules. The one that has gained the most traction with sedimentary geologists is that the curious little nodules formed mechanically, sort of like squeezing toothpaste. As the marsh plants rotted and produced methane gas, they created tubes and other hollow spaces. Subsequent compression by overlying layers of sediment squeezed the material below, which filled the open spaces and created nodules that look like scat. There is still debate, and more research is needed on these specimens, but we suspect that they will remain popular. Who wouldn't want such a curious object to show their friends?

As is happening at present in our modern warming world, the climatic changes in the Eocene played out not only on the land but also in the marine environment. (What is different now is the pace: at the modern rate of change, which is astronomic in comparison to change at the speed of geologic time, species are unable to adapt on a human time scale.) Forty million years ago, warm ocean tides lapped against the northwestern coast of North America. This was before the Cascade Mountains and long before the Olympic Mountains rose from the sea. What is now western Washington was a wide coastal plain with huge sluggish rivers that meandered from hundreds of miles inland and spilled into the sea. The dividing line between water and land co-incidentally extended north–south along a line roughly equivalent to Interstate 5. To the west lay the deeper-water seas and to the east were intertidal mudflats, brackish-water estuaries, and freshwater streams.

With warming ocean water came a reorganization of marine life, as animals that were better adapted to the new conditions moved in, took up residence, and evolved. No place in the state of Washington records this climatic change better than the Cowlitz Formation rocks of southwestern Washington. Over the past century, paleontologists have found an unrivaled collection of marine invertebrate fossils there. They have named almost two hundred species from these sandstones and siltstones, the highest marine fossil biodiversity in Washington, of which about 90% are mollusks—clams, snails, and scaphopods (tusk shells)—along with very rare nautiloids.

The most obvious fossil is the venerid (Venus) clam called *Pacificor*, a thick-shelled, ribbed, rotund clam that grew to 5 inches (13 cm) wide. Like most shallow-water clams, they burrowed in the loose sand below tide level and fed on particles filtered out of the seawater. Along the West Coast of North America, large Eocene venerid clams long had the generic name *Venericardia*, the same genus as more than a hundred

Slab of mudstone collected along the Cowlitz River, Lewis County, with many high-spired turret snails, *Turritella olequahensis*, plus a few clams (top right) and gray pebbles. The shells appear to be oriented in the same direction by wave action on the Eocene beach. The tops of the snails were worn away when the rock was exposed in the riverbank.

species found abundantly in Eocene sedimentary rocks in both the Gulf Coast and western Europe. *Venericardia* shells are so common in these rocks that nineteenth-century paleontologist Timothy Abbott Conrad described them as "that 'finger-post' of the Eocene . . . that may be regarded as the most characteristic fossil of its era" in *Descriptions of the Fossils and Shells Collected in California*.

In 2019, Argentinian paleontologist Damián Eduardo Pérez published a study of worldwide venerid clams that showed that earlier paleontologists had been too broad and had shoehorned too many species into the genus *Venericardia*. He determined that the name applies to the Eocene fossils only from France. Those from the Gulf Coast now have five different generic names, and those along the West Coast (Baja California to Washington) all belong to the genus *Pacificor*, a name published in 1953. Paleontologists had largely ignored that

This robust large Eocene Venus clamshell, *Pacificor clarki*, 5 inches (13 cm) in diameter, is from the Cowlitz Formation, Lewis County. Species of this genus are the marine index fossils from California to Washington for the more than 20-million-year Eocene Epoch.

This turret shell, *Turritella olequahensis*, 2.5 inches (6 cm) high, from the Cowlitz Formation shows predatory crab damage to the aperture (lower end of the shell).

An example of *Siphonalia sopenahensis*, 3 inches (7.6 cm) high, one of the commoner carnivorous snail shells in the Cowlitz Formation. This genus belongs in the family Buccinidae, which includes a wide variety of living snails called whelks.

proposed name, primarily because *Venericardia* was such a famous and comfortable name. *Pacificor* now has eighteen species.

The most common fossils in the Cowlitz sedimentary layers are the almost spherical moon snails in the family Naticidae, including the still living genera *Polinices*, *Neverita*, and *Sinum*. Moon snails fossilized very well, and thousands of adult and juvenile shells have been collected from the Cowlitz Formation. Based on the habits of modern moon snails, these extinct species were predators that located a potential meal by probing the sand and mud. They then enveloped their prey completely with their huge expanded muscular foot, employed a tongue-like rasp to drill a hole into the clamshell, and injected enzymes that dissolved the meat inside and allowed the moon snail to consume the liquefied flesh. This macabre feeding technique leaves a trace fossil of the distinctly beveled round drill hole in the prey shell. Eocene fossil shells with these distinct drill holes show that moon snails ate thin-shelled clams, other snails, and even other moon snails.

Other abundant marine species in the Cowlitz fauna are the high-spired turret shells, such as *Turritella olequahensis* and *Turritella weaveri*. Drill holes in their shells show that they were eaten by moon snails, too. In life, *Turritella* bury themselves almost horizontally in the sand and filter tiny food particles out of the seawater. They are very common in tropical sandy seashores today, heavily preyed on by fish, sea stars, carnivorous snails, and crabs, the latter peeling off the shell and extracting the meaty insides. Sometimes the crab is not successful and the snail can repair the broken shell. It is a rare *Turritella* shell, both modern and fossil, that shows no signs of predation. Other common Cowlitz fossils that suffered drill holes and crab ripping are the small conch-shell *Rimella washingtonensis* and the whelk *Siphonalia sopenahensis*, both of which probably ate marine worms.

As always happens in evolutionary arms races, snails evolved thicker and more ornate shells to protect themselves from crabs, which in turn evolved bigger and stronger claws. The most ornate shells in the Cowlitz deposits, with frills and spines, are the rock snails *Murex sopenahensis* and *Murex cowlitzensis*, though both species are rare. Their shells protected them from crab peeling and moon snail drilling. More common

A middle Eocene cone shell, *Conus vaderensis*, 2 inches (5 cm) long, from the Cowlitz Formation, is one of the oldest cone shells known. These marine snails are carnivorous, and some living species are poisonous.

This triton shell, *Cymatium cowlitzensis*, 4 inches (10 cm) high, is one of the larger snail shells from the Cowlitz Formation. Modern tritons are thick-shelled, live in tropical seas, and prey on sea urchins and other snails.

are the cone snails, *Conus vaderensis* and *Conus cowlitzensis*, which have smooth, thick shells with a narrow slit opening that made it hard for a crab to get a grip on the snail as it retreated within its shell. Many cone snails today have very toxic venom they use to kill or stun their prey. Other beautiful ornamented snails in this Eocene sea include species of the genera *Exilia*, *Turricula*, *Gemmula*, and *Cymatium*, all which live in tropical and semitropical seas today. Completing this world full of predators were sharks and rays, which left behind a few teeth.

The names of Cowlitz Formation fossil species reflect that these invertebrates were endemic to western Washington. Many of them were first described by paleontologist Charles E. Weaver. For example, Sopenah (*Siphonalia sopenahensis*) was the name of a Northern Pacific Railway stop by the Cowlitz River in Lewis County, previously called Little Falls; the town was later renamed Vader (*Conus vaderensis*), which sits between Olequa Creek (*Turritella olequahensis*) and the Cowlitz River (*Murex cowlitzensis*). Weaver, Washington State's first professional paleontologist, arrived at the University of Washington in 1907. Shy and reserved, he often taught from the back of the classroom, giving lectures that students described as very boring. In the summer field collecting trips, though, where he was in his outdoor element, his students reported that he was an enthusiastic teacher and had boundless energy. They were excited to be chosen as his field assistants even though Weaver did not drive and all field excursions were long hikes from the nearest railway line.

During his forty-three years at the University of Washington, Weaver mapped and described sedimentary rock units across Washington and collected Cenozoic marine fossils that formed the foundation of the Burke Museum's paleontology holdings. He spent three years in central Argentina collecting marine Cretaceous fossils and shipped tons of these back to Seattle. He is best known for his pioneering work with marine invertebrates, naming more than 150 fossil genera and species endemic to the Cenozoic of Washington and Oregon. For those who wish to know more, his three-volume book on the fossils of western Washington and Oregon, published in 1943, is still the most used text for identifying our Cenozoic fossil mollusks.

Weaver's work identifying and naming the many invertebrates that inhabited the Cenozoic coastal waters remains an essential scientific resource. His detailed shell descriptions and accepted classifications allow present-day paleontologists to determine the relationships between extinct and modern species and what that means for the paleoenvironment of our area. For example, if a modern clam is a cold-water dweller, this implies that an extinct relative would have similar environmental needs. If paleontologists didn't have these fundamental genera and species descriptions of the fossils, they would find it far harder to draw conclusions about the past.

The Cretaceous Period

145 Million Years Ago to 66 Million Years Ago

LASTING FROM 66 MILLION TO 145 MILLION YEARS AGO, the Cretaceous is the last period of the Mesozoic, a span of time famous primarily because of dinosaurs. But it was also when great marine reptiles, such as giant mosasaurs and long-necked elasmosaurs, roamed the oceans, along with another well-known group, the ammonites. Washington State, though, has very little Cretaceous sedimentary rock, which means we have very few fossils from this period.

This paucity of fossils led us to concentrate our stories on the beautifully preserved ammonites and a single dinosaur bone from Sucia Island, one of the San Juan Islands in northernmost Puget Sound. The fossils are found in Late Cretaceous fossiliferous marine rocks, which are part of a widespread sedimentary sequence called the Nanaimo Group. They extend along the east side of Vancouver Island and the Gulf Islands in British Columbia. (The cast of an elasmosaur, a very long-necked marine reptile that hangs above the stairway in the Burke Museum was found in Nanaimo rocks on Vancouver Island by a young girl fossil hunting with her father.)

There are other older Cretaceous marine fossiliferous units, but they are of limited extent. These rocks are widely scattered on the San Juan Islands and in the east side of the North Cascades, such as along the Nooksack River near Glacier, Washington, and in the Methow Valley. Paleontologists have collected both small and very large planispiral ammonites and many straight-shelled ammonite fossils from the Nooksack River. Most are only partial shells and are difficult to extract from the very dense, slightly metamorphosed siltstones and mudstones, which means, at this point in time, that they provide little scientific information and don't add to our stories of understanding Washington's geology and paleontology.

The Cretaceous was a time of high global temperatures and, as far as geologists can tell, there was little or no ice in the high polar regions, which enabled dinosaurs to live within the Arctic and Antarctic Circles. A lack of ice on the warm planet led to high sea levels, so high that a shallow Western Interior Seaway stretched from the Arctic Ocean to the Gulf of Mexico, cutting North America into two smaller continents. Because the Rocky Mountains first began to rise only in the Late Cretaceous, we assume that dinosaurs, and other animals, roamed across the landscape from Montana to the Washington coastline.

In the last 66 million years, sedimentary rocks that could have contained fossils of these animals have been squashed, faulted, and folded by the tectonic forces that shaped, and continue to shape, the western United States. In addition, in Washington the lava flows of the Columbia Plateau covered sites that could potentially have yielded fossils. At present, only a few isolated pockets of Cretaceous fossils occur in the state. This doesn't mean that others won't be found, but it does seem that for the time being, our story of the almost 80 million years of the Cretaceous will remain limited, though we do at least feature two of the best-known groups of extinct animals: ammonites and dinosaurs.

For as long as people have described ammonites, they have bestowed them with unique traits. Writing in the first century CE in his *Natural History*, Pliny the Elder observed that they were "among the most sacred gems of Ethiopia" with an ability that "ensures prophetic dreams." He was also the one who gave them their name, *Hammonis cornu*, a reference to Hammonis, or Ammon, the Egyptian sun god often depicted as a man with the head of a ram. Other properties attributed to ammonites included aiding hunters, treating stings, and providing relief to cramped cows, as well as increasing their milk production.

No one appears to have found such practical uses for Washington's ammonites, but while searching for them on Sucia Island, Jim Goedert and David Starr did locate what would become the state's first dinosaur. More than twenty species of ammonite have been found on the island, along with a single nautiloid in the genus *Eutrephoceras* and a variety of snails and clams, including the giant *Inoceramus*. Most of the fossils collected from Sucia are fragmentary and break easily as they erode out of the cliffs. However, there are magnificent complete specimens with the outer shell or the middle pearly layer of shell still intact.

Ammonites are a group of cephalopod mollusks that flourished in the Mesozoic seas and had geologically rapid species evolution and extinctions. They are the Mesozoic part of a larger group called Ammonoidea that first appeared in Devonian oceans some 400 million years ago. Incredibly diverse, with around 10,000 described species from all over the world, ammonites ranged from dime-size to more than 6.5 feet (2 m) in diameter. They went extinct at the same time as the dinosaurs (except for birds), 66 million years ago. Their living relatives include octopus, cuttlefish, squid, and the shelled pearly nautilus.

The majority of ammonites, like the common species from Sucia, are the classic cinnamon-roll shape, spiraling in one flat plane. Viewed from the front, most ammonites are rounded, but many are slim and

Canadoceras newberry-anum is found in western Washington, on Vancouver Island, and in southern Alaska. It has a strongly ribbed shell that varies in different localities in number and size of the ribs; this specimen is 5 inches (13 cm) in diameter.

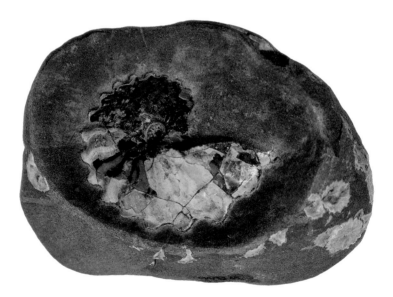

A complete juvenile *Hoplitoplacenticeras vancouverensis* was revealed when the concretion was split in half; the block is 8 inches (20 cm) wide. This species has short spines around the shell's outer edge that are usually broken during fossilization.

Two *Desmophyllites diphylloides* shells, each 3.5 inches (9 cm) in diameter, from the Late Cretaceous sedimentary rocks of Sucia Island. This tightly coiled, smooth-shelled species is the most abundant ammonite in the Sucia collections.

hydrodynamically streamlined. Several types have spines, which don't preserve very well, and others have low ribs running across the coils—for example, our local genera *Canadoceras*, *Eupachydiscus*, and *Hoplitoplacenticeras*. Some genera, including *Desmophyllites*, have the last coil covering up prior coils (involute), whereas in some the entire sequence can be seen from the side (evolute).

Ammonites also evolved more-complex shapes. Known as heteromorphs (meaning different forms), the shells appear to be in the process of uncoiling as the animal grew and spiraled outside the typical single flat plane. Heteromorphs can look similar to snail shells, like bent sticks, or have what appear to be wildly disorganized growth forms. Models (both physical and computerized) show that ammonites were not speedy swimmers. Most likely they floated within the middle depths of the oceans and must have grabbed whatever food items they could from the water or seafloor. Curiously, heteromorphic shell shape evolved separately in different ammonite families with different shell patterns, but researchers do not know why this happened.

Paleontologists have discovered a few heteromorph genera on Sucia Island, though usually in small pieces. The island's most common heteromorph, *Baculites inornatus*, has a minute initial whorl followed by a straight or slightly curved long shell, whereas the shells of *Pseu-*

Straight ammonite, *Baculites inornatus*, is the index species for the Late Cretaceous Cedar District Formation. This specimen may be a single shell that was broken before fossilization; the large section is 5 inches (13 cm) long. Together with the paleomagnetic signatures from the surrounding sedimentary layers, this species pins the age of the rocks at 80 million years ago.

doxybeloceras begin with a tight coil, proceed in a straight shell, and end in a big hook. Very common in the Late Cretaceous, some *Baculites* reached lengths of 6.5 feet (2 m). The name *Baculites*, which the great European zoologist Jean-Baptiste Lamarck first used in 1799, means "walking stick." In some sites around the world, this species is found in large numbers, suggesting that they lived (and died) together in large swimming schools, like some species of fish.

The Sucia ammonites are some of the many fossils within the off-shore mudstones of the Cedar District Formation. Deposited in the Late Cretaceous (80 million to 82 million years ago), the Cedar District sediments are part of a large suite of fossiliferous marine conglomerates, sandstones, and mudstones called the Nanaimo Group. These sediments are now exposed along Sucia and Orcas Islands, the Canadian Gulf Islands, and eastern Vancouver Island.

In addition to the diverse ammonites, fossils from the Nanaimo Group include a wide array of mollusks, crabs, and lobsters and a few scattered fish bones, as well as rare bird and marine reptile bones. Within the latter group are turtles, mosasaurs, and an exceptionally complete long-necked flippered plesiosaur. The bones of two elasmosaurid plesiosaurs, discovered thirty-two years apart in 1988 and 2020, are on display at the Courteney and District Museum, Vancouver Island. A completed cast made from these bones also hangs in the Burke Museum.

Lacking modern ammonites to study, researchers study the biology and behavior of their still-living relatives. All are characterized by a soft body with a mouth, surrounded by a large foot divided into arms or tentacles. Octopus have eight arms, squid have ten tentacles, and *Nautilus* have up to ninety small appendages, but there are no good impressions of ammonite soft bodies in the fossil record to tell us the number they had. Paleontologists hypothesize that ammonites probably had a small number of tentacles, as they are more closely related to squid than to the very large, but mostly extinct, Nautiloidea.

As in the nautiloids, the interior of the ammonite shell is divided into many short chambers sealed off by curved shell walls, called septa. The animal constructed the wall as it grew and lived attached to the inside of the most recently constructed, or ultimate, chamber. Unlike the simple curved chamber septa of *Nautilus*, those of ammonites were complex and serpentine where they attached to the outer shell wall. In ammonites, if the outer shell has worn away, one can see beautifully ornate patterns of the suture lines where they grew into the outer shell. The patterns are so distinct that paleontologists use them to assign specimens to specific genera and species.

The chambers were not completely shut off from the living animal. A tissue-lined thin tube, or siphuncle, ran inside the shell margin from the home chamber to the very first chamber. In modern *Nautilus*, which have a thicker shell than ammonites, the siphuncle enables the animal to change its buoyancy by slowly changing the chemical composition of the water within each chamber. This adaptation allows them to move slowly from great depth up to the surface to feed at night, where they

LEFT A modern *Nautilus* shell cut in half to show its cephalopod chamber and simple chamber-wall design.

RIGHT In contrast, this Sucia ammonite, *Pachydiscus* sp., shows very convoluted chamber-wall patterns, characteristic of Cretaceous ammonites.

find more food and fewer predators. Based on experiments applying pressures to model shells, paleontologists have concluded that the convoluted septal walls of ammonites compensated for their outer shell being much thinner than that of nautiloids. This would have allowed ammonites to safely live at depths and not have the overlying water pressure implode their shell and crush them.

Using their impressive sense of smell, large eyes, and big brain, living cephalopods are carnivores—either predators or scavengers of small prey items—and employ a hard pointed beak backed up in the mouth by a series of small "teeth" made of chitin. Based on what modern cephalopods eat, scientists propose that ammonites ate food ranging in size from plankton to much larger fare for the largest ammonites. They may even have been cannibalistic; ammonites produced thousands of tiny juveniles that floated as edible plankton in the upper parts of the ocean.

Because of the diversity and global distribution of ammonites, paleontologists have long been fascinated by their ecological success, distribution across time and space, and total extinction at the end of the Cretaceous. This extinction event occurred after a giant meteorite hit the Earth and dramatically degraded the global ecosystems with an instantaneous acidification of ocean surfaces. Paleontologists who have spent years studying ammonite distribution and biology have found that ammonite diversity declined toward the end of the Cretaceous. In addition, in the last half million years before meteor impact, most of the remaining thirty-one genera lived in small, restricted geographic areas. This seems to have made them vulnerable to extinction. A few widespread genera, including *Baculites*, however, are found in the sedimentary rocks deposited after the extinction event, but they lasted for only a few hundred thousand years. In contrast, many nautiloid genera survived the extinction event, including the Sucia Island *Eutrephoceras*, and continued to flourish for many millions of years.

Paleontologists have concluded that egg size may be the reason for the difference. Because many microfossils and groups that lived as plankton did not survive the impact, researchers hypothesize that ammonites produced millions of very tiny eggs, and the hatchlings must have floated as plankton on the top layers of the oceans. When the meteor hit and changed acid levels, the tiny juvenile ammonites were part of the global plankton that did not survive. Nautiloids, in contrast, produced a few large yolk-filled eggs, and the juveniles could have subsisted on this food source long enough to remain viable, survive, reproduce, and keep their species alive. Such are the vagaries of evolution and extinction.

It is illegal to collect fossils without a permit at Sucia Island Marine State Park. University of Washington paleontologists had a permit to collect with the proviso that the fossils become part of the Burke Museum collection. The Burke is the state's official natural history museum, thus all the fossils and biological specimens in its collection belong to the people of Washington.

Standing on a beach on Sucia Island, Brandon Peecook knew the second he saw the bone that it was from a dinosaur, the first ever discovered in Washington. It looked nothing like the classic image—big and easily identifiable—of dinosaur fossils as they are often portrayed in the popular press. Instead, the bone was barely discernible from the surrounding mudstone, particularly in the diffuse light of a cloudy May day. The only reason it had been noticed was that the discoverers, Jim Goedert and David Starr, had spent years exploring the fossil-rich bluff that rose above the boulder-strewn beach in the San Juan Islands.

The pair had been on one of their regular forays for ammonites when they noticed the embedded bone. Eroded to a smooth surface by waves, it stood out because of its size, about 18 inches (45 cm) long, oatmeal color instead of gray, and a spongy texture like the inside of a bone. Goedert took a photograph and sent it to the Burke Museum's curator of vertebrate paleontology, Christian Sidor. Because Goedert and Starr knew the rocks were 80 million years old, they had suspected that the fossil was either a dinosaur or a marine reptile that resembled the classic image of the Loch Ness monster.

Either way, the Burke paleontologists knew they had to go see the bone and figure out if they could bring it back to the museum. After taking the ferry to Orcas Island, Peecook and four other Burke researchers hitched a boat ride to Sucia with a ranger who worked at the island's Washington state park facility. He dropped them off on a kelp- and barnacle-covered fin of rock visible only at low tide, also the only time the fossil was easily accessible. Because of the tide, the group had a narrow window of about six hours to locate the fossil, determine what it was, and extract it. Their tools were rock hammers, mallets, and a saw with a diamond-tipped blade to cut into the hard mudstone.

The Burke team immediately realized the fossil had to be dinosaur because it had a hollow shaft, a feature not found in marine reptiles. They then began to hammer and saw around the bone, working in tighter circles to discern the edge of the fossil, and eventually cut away

Partial femur bone, 17 inches (43 cm) long, of a large carnivorous dinosaur collected from rocks of Sucia Island, San Juan County. The well-worn broken bone was hollow inside and filled with mud. Therapods, carnivorous dinosaurs, have characteristically hollow bones, and this bone was found in deep-water marine rocks, which indicates that at least part of the dinosaur carcass was washed from the land out to sea.

enough rock to leave the bone on a pedestal. By this time, the tide had covered most of the beach and they had only about forty minutes remaining until the ranger would return. It didn't leave them enough time to extract the bone without damaging it, so they decided to cut it and the surrounding material into two pieces, which meant they would lose a small bit of the fossil. Soon after, the ranger arrived and picked up them and the two pieces, which were the size of a shoe box and a banker box. Of the 150 feet (46 m) of beach that had been out of the water when they arrived, about 10 feet (3 m) remained.

Back at the Burke, Peecook's team put the two specimens in the basement, where they remained for the next eight months. During that period, Burke preparators slowly removed the excess rock when they had some free time. "The bone was interesting, but we had other work that took scientific priority," said Peecook. The Burke team ultimately glued together the two parts to reveal a dagger-shaped bone. It was 17 inches (43 cm) long, 9 inches (23 cm) wide at the top, and 1 inch (2.5 cm) wide at the bottom.

Now that he had the bone, Peecook began to try and figure out

Silhouette of a therapod dinosaur bone showing which part of the femur the Sucia Island bone is. The very large size shows that this bone was from a large dinosaur that lived 80–82 million years ago.

what it was. To do that required comparing the bone with specimens in other museums, which required travel money. As he began to seek funding, other members of the Burke community finally learned of the specimen and immediately provided the travel funds.

The discovery of the bone on Sucia Island came at a fortunate time. Over the past few decades, dinosaur paleontology has entered a golden age of discovery and research. The number of new species has skyrocketed as paleontologists have begun to work in areas such as Argentina, Mongolia, and China, which are outside the traditional dinosaur-rich locations of the western United States, Canada, and Europe. Paleontologists have also started to develop better understanding of the evolution, physiology, and ecology of dinosaurs, as well as many other extinct animals, by incorporating work from a diverse range of scientists, including ecologists, anatomists, engineers, and geologists.

By integrating these different fields, researchers are starting to bring dinosaurs to life, showing their complicated lives and lifestyles. For example, in contrast to how they had been portrayed by early generations of paleontologists, *Tyrannosaurus rex* were now understood to be agile, quick predators that ran with tails held horizontally, which acted like a balance beam. They had binocular vision and an excellent sense of smell, used their saw-like teeth for shredding, and matured quickly—gaining more than 100 pounds (220 kg) per month—if they survived their youth. Most didn't. They also sported feathers; some members of the tyrannosaurid family were covered completely by them, though they didn't fly.

At the Museum of the Rockies, in Bozeman, Montana, Peecook decided that the bone was either part of a femur from a hadrosaur, a group of plant-eating dinosaurs popularly called duckbills, or a tyrannosaurid, though it was clear that the Sucia specimen was not *T. rex*. Eventually, at the Royal Tyrrell Museum in Alberta, home to an extensive collection of dinosaur fossils, Peecook found what he called a dead ringer.

Daspletosaurus torosus were meat eaters in the tyrannosaurid family. Smaller than *T. rex*, they had larger teeth and lived prior to their more famous relative's reign. Evolving near the very end of the 170 million years of dinosaurs' existence, *T. rex* lived between 66 million and 68 million years ago. The Sucia Island dinosaur was not a *Daspletosaurus*, which have been found only in Alberta, Canada, but the similarity indicates that Washington's lone dinosaur was closely related to it.

Based on the location of the Sucia fossil in marine sediments, Peecook hypothesized that the dinosaur had perhaps died near a river and been washed into the sea, where it settled relatively close to a shoreline. What happened to the dinosaur's other body parts can never be known. They could have remained on land, never reached the sea, been scavenged and disbursed in the water, or washed away subsequent to ending up on the seafloor. Fortunately, one bone survived, providing a tantalizing view into the past.

The Paleozoic Era

539 Million Years Ago to 252 Million Years Ago

ALTHOUGH THIS FINAL CHAPTER of our book covers the longest amount of time, from 252 million to 539 million years ago, it contains the fewest number of Washington fossils. The Paleozoic was a time of great tectonic change across the globe. It was also period of great change in the planet's biota, with three mass extinctions, including the most devastating to the planet, which terminated this era. During what is called the Great Dying 252 million years ago, more than 95% of marine species and upward of 70% of terrestrial vertebrates died out.

These same large-scale changes are reflected in the rocks of Washington, in particular in the dearth of fossiliferous rocks we have from this time. At the beginning of the Paleozoic, sea levels were high and what is now the core of the North American continent was a landmass geologists call Laurentia. Not only was Laurentia much smaller than the modern North American continent, Laurentia was located on the equator, where it was joined by the smaller continent of Siberia, as well as Australia, then the northernmost tip of the supercontinent Gondwana. A warm circum-equatorial ocean current flowed past and around all the coastlines.

Before geologists had paleomagnetic signatures, and long before satellites and the Global Positioning System (GPS) allowed for more precision mapping, paleontologists had put together a map of the ancient continents based on the distribution of trilobites, an extremely diverse and widely spread group of marine invertebrates. The key that allowed paleontologists to do so was how trilobites from the tropical regions were very different from those that lived in the temperate regions and from those of cooler high latitudes. By matching the assemblages of fossil trilobite locations, they could create a map to show the location of ancient continents.

In Washington, our oldest Paleozoic fossils are marine invertebrates found in small isolated outcrops on the northern edge of Washington State, along the Idaho boundary. These early Cambrian to Ordovician fossils were deposited in sediments along the coastline of Laurentia, over a period of 100 million years. They are not, however, arranged tidily. Because the monumental forces of accreted terranes pushed assorted bits of rocks on top of and around the fossil-rich areas, we now see only small fossiliferous outcrops as "pop-ups" in a complicated geologic landscape.

Despite the paucity of fossil-rich layers of Paleozoic rocks, we are fortunate to have a relatively diverse array of trilobites. In this chapter, we focus on two middle Cambrian localities in eastern Washington. The better-known site is a very large quarry (on private property) near the town of Metaline, Pend Oreille County. Sadly, the second site comes from a temporary roadcut now backfilled and unavailable, but the fossils are beautifully preserved.

We complete our story of Paleozoic fossils with some of the oldest animals on Earth, the archaeocyaths. Early sponge relatives, they built the planet's first biological reefs. Cambrian fossils show the major evolutionary events when animals arose and diversified into every phylum that we have on Earth today.

Trilobites and archaeocyaths are not the lone Paleozoic fossils in the state. Throughout the rest of the Paleozoic, up to 252 million years ago, plate tectonics continued to move the continents, eventually assembling them into the single supercontinent called Pangaea. Laurentia was now attached to other landmasses but remained spanning the equator. In Washington, the representatives of these traveling landmasses are slivers of marine sedimentary rocks within different accreted terranes, which had traveled great distances before being slammed into the Laurentian continent. Because of how they are assembled, these little slices of rocks—the pop-ups that survived the tectonic chaos—are completely out of their original order, in the Okanogan Highlands, the North Cascades, and the San Juan Islands. All the fossils that have been found date from the Silurian to the Permian and are rare and in bad shape. They include ancient clams and snails, very rare ammonites, corals, and strange conical shells called hyoliths, as well as the broken stalks (columnals) of large crinoids. Unfortunately, the age of each of these deposits remains a scientific work in progress, and we cannot yet provide a coherent account of their story.

Trilobites were Earth's first charismatic animals and are arguably second on the list of best-known fossils, after the far scarcer dinosaurs. During their 270 million years of existence, from 252 million to 520 million years ago, trilobites evolved into at least 25,000 species ranging from a pinhead in size to as long as a tennis racket. Trilobites first appear in the fossil record 20 million years after the beginning of the Cambrian. Cambrian trilobites lived on the seafloor, scurrying across its substrate searching the shallow, wave-washed depths for tiny pieces of food, swimming throughout the water column, or floating on the surface. They diversified rapidly into many different families and spread around the world into different marine environments on the seafloor or floating in the ocean.

Like crabs, trilobites are marine arthropods covered in a hard calcium-carbonate exoskeleton. Their bodies consisted of three longitudinal lobes (hence their name), some elaborately spined and horned, others streamlined. They also had three body sections—tail, thorax, and head—and most had large eyes. Unlike the eyes of almost all other animals, the trilobites' eye lenses were made of calcite crystals instead of protein, which gave them what some call a stony stare and allowed the lenses to fossilize, which happens only rarely with other animals. Their eyes were also extraordinarily complex, especially for an animal that evolved so early. Covered by a single cornea, each eye consisted of hundreds or even thousands of individual lenses, like the eye of a bee. The lenses were closely packed in a hexagonal pattern and the whole set was arranged in a crescent shape to give almost 360-degree vision. Each lens was composed of a single calcite crystal—which provided great visual acuity—and acted independently to create a mosaic image that allowed the trilobite to distinguish shapes, light and dark zones, and other organisms on the seafloor. This was particularly important because as early as the middle Cambrian, the

Drawing of trilobites living on the seafloor.
Artwork by Heinrich Harder, 1916.

trilobite world was beset with small and large arthropod predators that also had excellent compound eyes, grasping appendages, and biting mouths. Seeing well was important.

Like beetles, trilobites molted their hard coverings as they grew, and these molts, whether whole or broken, are the most abundant fossils; a single animal has the potential to molt many times, creating many potential fossils. The molted exoskeletons of trilobites do not include the front of the head with the complex eyes, but they are still identifiable to species. Nor do many trilobite fossils include the antennae and the legs. Rare cases from sites with exceptional preservation, however, do show parts of the trilobite underside with the thorax made of segments, each having a pair of legs and a pair of attached gills. These little jointed legs (like shrimp's legs) cannot be seen from the top of the animal, but they did leave a variety of traces fossilized on the rock surface from walking, scratching at the seafloor, plowing through the top of the mud, or resting. Some smaller species also floated in the water column and could have traveled huge distances across the oceans.

Based on the similarity between the mouthparts of trilobites and crabs, paleontologists assume that most trilobites ate tiny pieces of organic detritus off the seafloor and that some were scavengers or even predators on other invertebrate animals on the seafloor. They were also prey items for larger arthropods, such as sea scorpions, nautiloids, and, later in the Paleozoic, predatory fishes.

One of the state's best-known trilobites is *Ogygopsis klotzi*, which grew to 5 inches (12 cm) long. The species is noted for a bulge in the middle of its head—filled not with brains but with a stomach—and its lack of spines. Their fame, though, has less to do with our specimens than with ones found in the Mount Stephen trilobite fossil beds of the famous Burgess Shale, in the Rocky Mountains of eastern British Columbia. The Burgess Shale includes layers with exceptionally well-preserved,

Ogygopsis klotzi, the most commonly found trilobite, preserved in the dark limy mudstone of the Metaline Formation, with a penny for scale. This very well-preserved specimen from the Canadian Rocky Mountains shows a compact body divided into three parts.

unusual, intact, and often soft-bodied marine invertebrates, which have been central in understanding the stories of early life on Earth. In contrast, Washington's *Ogygopsis* specimens are not terribly well preserved or whole; instead, paleontologists most often find only bits and pieces, and many of the fossils are distorted.

Fossils such as *Ogygopsis* come from the state's premier trilobite locality, the Metaline Formation. A thick sequence of dark gray limestones and shales, the formation is middle Cambrian to early in the Ordovician. Kenneth McLaughlin and Betty Enbysk of Washington State University published the first significant study of these fossils, in 1950. The fossils occur in the numerous layers of the huge, and now closed, Lehigh Portland Cement quarry, just east of town. For more than a century,

A small *Glossopleura* species from a road construction site in an iron-rich (rust-colored) mudstone. This 0.8 inch (2 cm) long specimen shows the body divisions of the head on the right side, the segmented thorax in the middle, and the unsegmented tail section to the left. Well-preserved *Glossopleura* specimens have long backward-pointing spines on both sides of the head.

other quarries in the area produced metals such as silver, gold, lead, and zinc, which gave the town its name.

More recently, avocational paleontologists, including Glen Schofield from Spokane, are collecting a diversity of trilobite species from the different layers within the quarry. In addition to *Ogygopsis klotzi*, twelve trilobite species in the genera *Elrathina*, *Poliella*, *Chancia*, *Fieldaspis*, and *Yohoaspis* have been identified. Other fossils found in the Metaline rocks include a wide variety of sponges, very early relatives of crinoids, rare brachiopods, and primitive mollusks. These trilobite genera are all present in the Burgess Shale, which has a date of 508 million years old, but the Metaline species are a bit younger and correspond with 506-million-year-old fossils found in Nevada and Utah. However, the top of the Metaline Formation is directly underlying or even interfingering with the overlying Ledbetter Shale, which is 485 million to 490 million years old, indicating that the sediments exposed in this quarry accumulated over a very long time.

There are two other local trilobite localities, one collected by paleontologist Linda McCollum from a small construction project on military property near Cheney. This site includes the middle Cambrian trilobites *Zacanthoides libertyensis*, *Glossopleura boccar*, and *Ehmaniella* sp. From her studies of Cambrian trilobites in Nevada, McCollum determined that these trilobites from Spokane County are around 503 million years old. Both the Washington trilobite sites described above, as well as the Canadian Burgess Shale, are within middle Cambrian times, within the Miaolingian Epoch, named after exceptional fossil sites in the Miaoling Mountains, southern Guizhou Province, China.

The other trilobite-bearing rocks in northeastern Washington are

A slab of rock with numerous partial olenellid trilobites from the Eager Formation in British Columbia, the same age as the Addy Formation in Washington.

the older quartz-rich red sandstone layers of the Addy Formation. Addy rocks crop out south of Colville, in Stevens County, and the trilobites *Nevadella addyensis* and *Esmeraldina argenta* are index fossils for the time period 516 million to 518 million years ago. Other body fossils from the Addy sedimentary rocks are all rare and include brachiopods and small conical shells called hyoliths. There are a wide variety of trace fossils in the Addy layers, illustrating that the seafloor teemed with different animals that did not leave body fossils. These traces include distinct scratch marks from the multi-legged trilobites walking through the mud and feeding traces of wormlike invertebrates.

As the name suggests, *Nevadella* were originally found in Nevada and California in rocks that could be very precisely dated by radiometric methods. *Esmeraldina* also occur in these rocks, and both genera are related to the common genus *Olenellus*, all within the most primitive order of trilobites, the Redlichiidae. Distinguishing features of these Cambrian trilobites include large eyes, many segments to the thorax that end in backward-pointing spines, and a very tiny tail end (pygidium).

If it seems remarkable that paleontologists can give such a specific and narrow time span in such ancient rocks, it's because trilobites very rapidly evolved and became extinct. There are 180 described families of trilobites, some with members that persisted for tens of millions of years and some that lasted only a few million years. The greatest diversity was in the late Cambrian and the Ordovician, with species from 63 families living at the same time. Because of trilobites' rapid changes, many of their remains have become index fossils that provide paleontologists with a window into one of the Earth's oldest animals.

In a 1920 report on the geology of Stevens County, pioneering mapping geologist and paleontologist Charles E. Weaver wrote that the "total absence of fossils" from the area's oldest rocks "renders [their] separation into formations of definite geologic age almost impossible." By 1936, however, geologists could report on forty-four fossil-bearing localities in the county and adjacent Pend Oreille County. With these fossils, geologists were able to designate the rocks as the oldest fossiliferous rocks in the state, placing them in the lower Cambrian. Other archaeocyaths have been found in isolated locations in northern Stevens County and in neighboring British Columbia rocks.

First described and named in 1861, *Archaeocyathus* perplexed paleontologists because of the animals' "extraordinary and interesting combination of characters," wrote paleontologist Fielding Meek in 1868. He initially thought they were corals, then switched to foraminifera. Other researchers vacillated between plants, green algae, sponges, and corals until the 1920s, when the "sponge school" began to dominate. Archaeocyaths are now listed as an extinct class within the phylum Porifera, along with all sponges. (Archaeocyath means "ancient sponge," though they have also been called *Cyathospongia*, meaning "sponge-sponge.") In contrast to sponges, archaeocyaths do not have the characteristic feature of a mesh of microscopic calcite or silica spines within their tissue, though they do resemble some modern sponges that live in deep-sea caves.

The first paleontologist to conduct extensive studies of archaeocyaths from Washington and southern British Columbia was Canadian paleontologist Vladimir Okulitch. He was one of the world's experts on archaeocyaths and literally wrote the book on archaeocyath taxonomy. In 1958, he and his doctoral student Robert Greggs listed ten archaeocyath genera from one hillside outcrop of the Reeves Limestone (previously called the Old Dominion Limestone) near Colville, in

Diagram of archaeocyath cup.

Stevens County. Since then, no other studies of these earliest animals in Washington have been published, and this site is no longer accessible.

The oldest fossil archaeocyaths are about 530 million years old, from very early in the Cambrian, making them among the oldest animals in the world. Abundant in the warm shallow seas of all Cambrian continents, these small calcareous sponges rapidly evolved into a high diversity of genera and species. Then they became extinct around 510 million years ago, with just a couple of genera hanging on for a few million years more. Archaeocyaths are the index fossil for lower Cambrian rocks globally.

Rising up 1–2 inches (2.5–5 cm) above the substrate, archaeocyaths were usually conical, occasionally cup-shaped, and more rarely spread into low mounds. In order to identify the fossils, researchers use high-power microscopes and look at thin petrographic sections. Many archaeocyaths had a porous double-wall structure and an array of side walls (called septa) that trapped tiny particles of food carried by water flowing through the pores. The animals lived attached to the lime-sediment-covered seafloor in shallow water. Their hard parts were also composed of calcium carbonate, the main ingredient in lime. Each cone or cup was less than 1 inch (2.54 cm) in diameter. If the animals and sediments became a rock—a limestone—it is very difficult to separate the fossil from its surrounding matrix, which has resulted in many archaeocyaths being studied in cross section.

Archaeocyaths lived in the shallow seas with other primitive invertebrates, including true sponges, brachiopods, primitive relatives of sea stars, unusual mollusks, and trilobites. The limy sediments in these shallow, warm seas were composed of trillions of tiny calcium-carbonate particles shed by dense mats of bacteria, also known as blue-green algae, colloquially called pond scum. Because the bacteria photosynthesized,

Archaeocyath reef limestone slab from Stevens County is 6 inches (15 cm) wide. The broken sponge fossils are gray calcium carbonate and the limestone rock is russet red from iron staining of the original lime mud. These archaeocyaths have not been identified, and the slab may contain more than one species.

Close-up view of the archae-ocyath cups from the slab above. The round gray fossil in the middle is a cross section of a single archaeocyath cup, 0.4 inch (1 cm) wide, showing the septa walls radiating around the outer rim. Above that is a longitudinal section of a cup showing the flaring top of the sponge and hollow central cavity.

the seawater would have been clean and clear—the perfect place for the filter-feeding archaeocyaths. In some places, they became the Earth's first animal reef builders, and together with the photosynthesizing bacterial mats, they would have stabilized the soft mud and provided hard surfaces for other organisms to settle, as occurs in modern coral reefs. The archaeocyath reefs were home to other newly evolved primitive invertebrates and, most likely, refuges for newly hatched juveniles of these early animals. In contrast to modern reefs, however, these small archaeocyath-algal reefs were built without corals, which didn't evolve until late in the Cambrian.

Paleontologists are always interested in the origin and extinction of organisms and in searching for the probable cause of the extinctions. The voluminous increase in the study of ancient rocks, aided by new techniques in geochemical investigations, has opened up an interesting world for paleontologists to look at how plants and animals lived and what forces changed them. Such studies have brought a new perspective on the limited geologic story of archaeocyaths. Twenty million years is a short history for an entire invertebrate group, compared with the hundreds of millions of years that trilobites and ammonites existed.

One of the great events in geologic history is the so-called Cambrian Explosion, when animals appeared. They developed eyes and brains. They began to hunt each other. They evolved into the major groups still in existence. But what is an animal? The kingdom Animalia contains all those organisms that are multicellular (compared with single-celled bacteria or amoeba) that do not make their own food (as plants do) and that can move independently, at least in one of the life-cycle phases. Animals developed muscles very early in their evolution, and since we are animals, this is the most easily understood major taxonomic kingdom. Geologists have established the start of the Cambrian not on an animal-body fossil but on the first large burrows that go down vertically into the sediment. These distinctive burrows called *Treptichnus pedum* are recognized all over the world.

We now know that "explosion" is not the right word for this expansion of life across all the shallow seas on Earth, because it took 40 million to 50 million years. The first animals evolved from an unknown single-celled ancestor before the Cambrian, in the Ediacaran Period (539 million to 635 million years ago).

The earliest animal phylum to evolve are true sponges, and they appear in the fossil record before the Cambrian boundary. But they are scattered and not useful time markers. However, their presence does show that animals did evolve early on. There are numerous other organisms (simply called small shelly fossils) in the Precambrian fossils record and, notably, the Ediacaran suite of multicellular organisms attached to the ground, dating from 560 million years ago. However, paleontologists are not sure what these creatures are and where they belong in the tree of life.

The early Ediacaran history was tied to the amount of oxygen available in the ocean waters. Unlike the vast majority of time that animals and plants have been on Earth, the open oceans of the Ediacaran had almost no dissolved oxygen, called anoxic seas. Around the time of the Ediacaran-Cambrian boundary and into the Cambrian, the shallow seas experienced frequent fluctuations in dissolved oxygen levels that greatly influenced life. Geochemists have identified eight worldwide cycles in the Cambrian when oxygen levels increased in the shallow shelf seas where the animals lived, interspersed with anoxic periods. The oxygen levels were less than 25% of what the oceans carry today. These fluctuations greatly affected these new animals. Two of the larger oxygen floods coincided with the expansion of archaeocyaths species at 524 million years ago. Trilobites first appeared about 3 million years later, with the next oxygen pulse. The other invertebrate groups appeared in the fossil record at different times during periods of oxygen pulses. A major extinction, the Sinsk Event, was the result of large anoxic floods onto the continental edges, which persisted for more than 2 million years. At that time, almost all archaeocyaths went extinct and very few trilobites survived. It took the animals a long time to reestablish themselves after that.

Gradually through the early and middle Cambrian, all the other animal phyla evolved, including the earliest fish, the ancestral group to all other vertebrates. It is apparent from detailed studies of lower Cambrian sedimentary layers, particularly in Siberia and China, that animals did not suddenly evolve all at once, but they have a long history before trilobites appeared.

As these oxygen fluctuations impacted life so drastically, we can now say that there was no single Cambrian explosion of life on Earth. Organisms evolved, changed, and became extinct from late Ediacaran times, and new animal groups evolved through the Cambrian and Early Ordovician. Paleontologist Rachel Wood explained the evolutionary changes this way: "For a long time, it has been thought that the Cambrian Explosion was an event, when all the major groups of animals appeared in the geological record over only a few million years. But now Cambrian-type fossils have been found in older Precambrian rocks, suggesting that the 'Cambrian Explosion' in fact had a deeper, older root—in other words, a longer fuse."

ACKNOWLEDGMENTS

WE WOULD LIKE TO THANK the following for their research, discussions, and inspiration (listed roughly in the order in which they appear in the book):

Ron Eng and Katie Anderson, Paleontology Collections managers at the Burke Museum, who cheerfully provided so much help for this book

Bruce Crowley, Burke Museum fossil preparator (retired), whose work has been essential to many of the beautiful specimens in the museum collections

Jim Goedert, Burke Museum research associate, for his wealth of information on Cenozoic marine fossils of the Pacific Northwest

Ruth Martin, Maureen Carlisle, Don Hopkins, and Paul Kester, Burke Museum paleontologists who shared their expertise and enthusiasm

George Mustoe, Western Washington State University, fossil trackways, trace fossils, and all aspects of paleobotany

Eric Gustafson, University of Oregon, the Ringold fauna

Patrick Lubinski, Central Washington State University, Wenas mammoth

Robyn Burnham, University of Michigan, and Renee Love, University of Idaho, Eocene leaf fossils

Melanie Devore, Georgia College and State University, and Kathleen Pigg, Arizona State University, for sharing their paleobotany passion and expertise

Kirk Johnson, Sant Director of the Smithsonian National Museum of Natural History, for his botany expertise and Stonerose interest

Don Grayson, University of Washington, Quaternary mammals

Steve Kenady, Cascadia Archaeology, and Michael Wilson, Douglas College, New Westminster, British Columbia, Ayer Pond bison and Orcas Island sloth

Cathy Whitlock, Regents Professor at Montana State University, Quaternary pollen

Elizabeth Wheeler, North Carolina State University, and T. A. Dill-hoff, Evolving Earth Foundation, for their work on fossil woods

Rod Feldmann and Carrie Schweitzer, Kent State University for all things crustacean

Damián Eduardo Pérez, Instituto Patagonico de Geologia y Paleontologia, Eocene bivalves

Larry Workman and the Quinault Indian Nation, for welcoming us on their land

Carlos Mauricio Peredo, Miami University; Nicholas Pyenson, Smithsonian National Museum of Natural History; and Mark Uhen and Margot Nelson, George Mason University, cetaceans living and fossil

Jack Tseng, University of California, Berkeley, fossil mammals

Carole Hickman, University of California, Berkeley, my graduate student supervisor and for Pacific Northwest molluscan paleontology

Kathleen Campbell, University of Auckland, for all her experience with hydrocarbon seeps

Linda McCollum and Ernie Gilmour, Eastern Washington University, and Fred Sundberg, Museum of Northern Arizona, Cambrian faunas

Glen Schofield, Spokane, avocational paleontologist, trilobites

Rachel Wood, University of Edinburgh, Scotland, rise of animals in the Cambrian

WE WOULD ALSO LIKE TO ACKNOWLEDGE the contributions of collectors and researchers from the past whom we did not name or whose backgrounds we did not provide in the text (also listed roughly in the order in which they appear in the book):

Kenneth McLaughlin and Betty Enbysk, Washington State University, the first to describe the Metaline trilobites. Enbysk was the first woman to graduate with bachelor's and master's degrees from WSU; she later received her PhD from the University of Washington focusing on foraminifera and had a long career in oceanography.

Nellie May Tegland, the first to thoroughly study the Blakeley Formation fossils. She died at the age of forty-three, just months after receiving her PhD from University of California, Berkeley. Six years earlier, she had gotten her master's degrees at the University of Washington. Little is known about her beyond her contributions to paleontology.

Wes Wehr, Republic fossils. Beginning in the 1970s, his interest in these fossils was essential to their scientific and popular rebirth. A man of many talents—painter, composer, author, and affiliate curator of paleobotany at the Burke Museum—Wehr continued to research, collect, promote, and publish papers about the Republic fossils until he died in 2004.

Gordon Simmons, Stan Mallory, and Robert Greengo, Burke Museum, Sea-Tac sloth

George B. and Ruth L. Peabody, and Haakon and Aslang Friele, Blue Lake rhino

Charles E. Weaver, University of Washington, Burke Museum collections and molluscan taxonomy

Jack Wolfe, United States Geological Survey, paleobotany of Puget Group leaves

AND FINALLY, we would also like to thank the UW Press for bringing our book to life: Nicole Mitchell for her support from the beginning; Andrew Berzanskis for his advice and support; Kris Fulsaas for her wonderful copyediting; and Jennifer Comeau, Mindy Basinger Hill, and Larin McLaughlin for their help in design and production. As they say, it takes a village.

REFERENCES

General Reading

Johnson, Kirk, and Ray Troll. *Cruisin' the Fossil Coastline: The Travels of an Artist and a Scientist along the Shores of the Prehistoric Pacific*. Golden, CO: Chicago Review Press/Fulcrum, 2018.

Ludvigsen, Rolf, ed. *Life in Stone: A Natural History of British Columbia's Fossils*. Vancouver: University of British Columbia Press, 1996.

Ludvigsen, Rolf, and Graham Beard. *West Coast Fossils: A Guide to the Ancient Life of Vancouver Island*. Rev. ed. Madeira Park, BC: Harbour Publishing, 1998.

Miller, Marli B., and Darrel S. Cowen. *Roadside Geology of Washington*. 2d ed. Missoula, MT: Mountain Press, 2017.

Orr, Elizabeth L., and William N. Orr. *Oregon Fossils*. 2d ed. Corvallis, OR: Oregon State University Press, 2009.

Tucker, David. *Geology Underfoot in Western Washington*. Missoula, MT: Mountain Press, 2015.

Williams, Hill. *The Restless Northwest: A Geological Story*. Pullman: Washington State University Press, 2002.

Profiles

1 CHARISMATIC MEGAFAUNA MAMMOTHS

Grayson, Donald K. *Giant Sloths and Sabertooth Cats: Extinct Mammals and the Archaeology of the Ice Age Great Basin*. Salt Lake City: University of Utah Press, 2016.

Lubinski, Patrick M., James Feathers, and Karl Lillquist. "Single-Grain Luminescence Dating of Sediment Surrounding a Possible Late Pleistocene Artifact from the Wenas Creek Mammoth Site, Pacific Northwest, USA." *Geoarchaeology: An International Journal* 29 (2014): 16–32.

Nisbet, Jack. *Visible Bones: Journeys across Time in the Columbia River Country*. Seattle: Sasquatch Books, 2004.

Grayson, Donald K., and David J. Meltzer. "Revisiting Paleoindian Exploitation of Extinct North American Mammals." *Journal of Archaeological Science* 56 (2015): 177–93.

Haynes, C. V. Jr, and B. B. Huckell. "The Manis Mastodon: An Alternative Interpretation," *PaleoAmerica* 2 (2011): 189–191.

Hoppe, Kathryn A., and Paul L. Koch. "Reconstructing the Migration Patterns of Late Pleistocene Mammals from Northern Florida, USA." *Quaternary Research* 68 (2007): 347–352.

Waters, Michael R., Zachary A. Newell, H. Gregory McDonald, Daniel C. Fisher, Jiwan Han, Michael Moreno, and Andrew Robbins. "Late Pleistocene Osseous Projectile Point from the Manis Site, Washington—Mastodon Hunting in the Pacific Northwest 13,900 Years Ago." *Science Advances* 9 (February 1, 2023).

3 THE SEA-TAC SLOTH

Kenady, Stephen M., Michael C. Wilson, Randell F. Schalk, and Robert R. Mierendorf. "Late Pleistocene Butchered *Bison antiquus* from Ayer Pond, Orcas Island, Pacific Northwest: Age Confirmation and Taphonomy." *Quaternary International* 233 (2011): 130–141.

Mann, Daniel H., Pamela Groves, Benjamin V. Gaglioti, and Beth A. Shapiro. "Climate-Driven Ecological Stability as a Globally Shared Cause of Late Quaternary Megafaunal Extinctions: The Plaids and Stripes Hypothesis." *Biological Reviews* 94 (2019): 328–352.

McDonald, H. Gregory. "The Sloth, the President, and the Airport." *Washington Geology* 26, no. 1 (1998): 40–42.

Williams, David B. "Construction at Sea-Tac Airport Unearths an Extinct Giant Sloth on February 14, 1961." *HistoryLink*, April 24, 2010, essay 9408. www.historylink.org/File/9408.

4 WHAT WE LEARN FROM POLLEN

Barnosky (Whitlock), Cathy. "Late Quaternary Vegetation near Battle Ground Lake, Southern Puget Trough, Washington." *Geological Society of America Bulletin* 96 (1985): 263–271.

Leopold, Estella B., Peter W. Dunwiddie, Cathy Whitlock, Rudy Nickmann, and William A. Watts. "Postglacial Vegetation History of Orcas Island, Northwestern Washington." *Quaternary Research* 85 (2016): 380–390.

Porter, Stephen C. "Glaciation of Western Washington, USA." *Development in Quaternary Sciences* 15 (2011): 531–535.

Williams, David B. "Leopold, Estella (b. 1927)." *HistoryLink*, March 27, 2010, essay 9378. https://historylink.org/File/9378.

5 A MILLION-YEAR-OLD MIGRATION

Smith, Gerald R., David R. Montgomery, N. Phil Peterson, and Bruce Crowley. "Spawning Sockeye Salmon Fossils in Pleistocene Lake Beds of Skokomish Valley, Washington." *Quaternary Research* 68 (2007): 227–238.

Wilson, Mark V. H., and Guo-Qing Li. "Osteology and Systematic Position Salmonid *Eosalmo driftwoodensis* Western North America." *Zoological Journal of the Linnean Society* 125 (1999): 279–311.

6 WASHINGTON'S FIRST DEER

Fry, Willis E., and Eric P. Gustafson. "Cervids from the Pliocene and Pleistocene of Central Washington." *Journal of Paleontology* 48 (1974): 375–386.

Gustafson, Eric P. "Early Pliocene North American Deer: *Bretzia pseudalces*, Its Osteology, Biology and Place in Cervid History." *Bulletin of the Museum of Natural History, University of Oregon* 25 (2015).

Smith, Gerald R., Neil Morgan, and Erick Gustafson. "Fishes of the Mio-Pliocene Ringold Formation, Washington: Pliocene Capture of the Snake River by the Columbia River." *University of Michigan Papers on Paleontology* no. 32 (2000).

7 THE STONE RHINO

Chappell, Walter M., J. Wyatt Durham, and Donald E. Savage. "Mold of a Rhinoceros in Basalt, Lower Grand Coulee, Washington." *Geological Society of America Bulletin* 62 (1951): 907–918.

Kasbohm, Jennifer, and Blair Schoene. "Rapid Eruption of the Columbia River Flood Basalts and Correlation with the Mid-Miocene Climate Optimum." *Science Advances* 4 (September 19, 2018).

Maguire, Kaitlin C., Joshua X. Samuels, and Mark D. Schmitz. "The Fauna and Chronostratigraphy of the Middle Miocene Mascall Type Area, John Day Basin, Oregon, USA." *PaleoBios* 35 (2018): 11–51.

Williams, David B. "Climbers Find Basalt Mold and Bones of a 15-Million-Year-Old Rhinoceros at Blue Lake, Grant County, in July 1935." *HistoryLink*, May 5, 2010, essay 9409. https://historylink.org/File/9409.

8 WASHINGTON'S MOST BEAUTIFUL FORESTS

Beck, George F. "Exotic Ancient Forests of Washington." *Northwest Science* 10 (1936): 22–24.

Mustoe, George E. "Washington's Fossil Forests," *Washington Geology* 29, no. 2 (2001): 10–21.

Wheeler, Elisabeth A., and Thomas A. Dillhoff. "The Middle Miocene Wood Flora of Vantage, Washington, USA." *International Association of Wood Anatomists IAWA Journal* Supplement 7 (2009).

9 TRACKING THE TERRIFYING BIRDS

Mustoe, George E., Richard M. Dillhoff, and Thomas A. Dillhoff. "Geology and Paleontology of the Early Tertiary Chuckanut Formation." In *Floods, Faults, and Fire: Geological Field Trips in Washington State and Southwest British Columbia*, 121–135. Vol. 9 of *Geologic Society of America Field Guide*. Boulder, CO: GeoScienceWorld, 2007.

Mustoe, George E., David Tucker, and Keith L. Kemplin. "Giant Eocene Bird Footprints from Northwest Washington, USA." *Palaeontology* 55 (2012): 1293–1305.

10 FUELING THE ECONOMY WITH FOSSIL FLORA

Breedlovestrout, Renee R., Bradly J. Evraets, and Judith Totman Parrish. "New Paleogene Paleoclimate Analysis of Western Washington Using Physiognomic Characteristics from Fossil Leaves." *Palaeogeography, Palaeoclimatology, Palaeoecology* 392 (2013): 22–40.

Burnham, Robyn J. "Paleoecological and Floristic Heterogeneity in the Plant-Fossil Record: An Analysis Based on the Eocene of Washington." *US Geological Survey Bulletin* 2085-B (1994): 1–36.

Hay, O. P. "Descriptions of Two New Species of Tortoises from the Tertiary of the United States." *Proceedings, US National Museum* 22, no. 1181 (1889): 21–28.

Wolfe, Jack A. *Paleogene Biostratigraphy of the Nonmarine Rocks in King County, Washington.* US Geological Survey Professional Paper 571, 1968.

11 STONEROSE

Greenwood, David R., Kathleen B. Pigg, James F. Basinger, and Melanie L. DeVore. "A Review of Paleobotanical Studies of the Early Eocene Okanagan (Okanogan) Highland Floras of British Columbia, Canada, and Washington, USA." *Canadian Journal of Earth Sciences* 53 (2016): 548–564.

"Republic Centennial Issue." *Washington Geology* 24, no. 2 (1996): 2–44.

Wehr, Wesley C. "Early Tertiary Flowers, Fruits and Seeds of Washington State and Adjacent Areas." *Washington Geology* 23, no. 3 (1995): 3–16.

Wehr, Wesley C., and Donald Q. Hopkins. "The Eocene Orchards and Gardens of Republic, Washington." *Washington Geology* 22, no. 3 (1994): 27–34.

12 THE LAST NAUTILOIDS

Goedert, James L., and Steffen Kiel. "A Lower Jaw of the Nautiloid *Aturia angustata* (Conrad, 1849) from Oligocene Cold Seep Limestone, Washington State, USA." *PaleoBios* 33 (2016): 1–6.

Schweitzer, Carrie E., and Rodney Feldmann. "Fossil Decapod Crustaceans from the Late Oligocene and Early Miocene Pysht Formation and Late Eocene Quimper Sandstone, Olympic Peninsula, Washington." *Annals of Carnegie Museum* 68 (1999): 215–273.

Ward, Peter D., Frederick Dooley, and Gregory J. Barord. "Nautilus: Biology, Systematics, and Paleobiology as Viewed from 2015." *Swiss Journal of Palaeontology* 19 (February 2016).

13 CLAMS AND BACTERIA

Campbell, Kathleen A. "Recognition of a Mio-Pliocene Cold Seep Setting from the Northeast Pacific Convergent Margin, Washington, USA." *Palaios* 7 (1992): 422–433.

——. "Hydrocarbon Seep and Hydrothermal Vent Paleoenvironments and Paleontology: Past Developments and Future Research Directions." *Palaeogeography, Palaeoclimatology, Palaeoecology* 232 (2006): 362–407.

Hickman, Carole S. "Composition, Structure, Ecology and Evolution of Six Deep-Water Mollusk Communities." *Journal of Paleontology* 58 (1984): 1215–1234.

Levin, Lisa A. "Ecology of Cold Seep Sediments: Interactions of Fauna with Flow, Chemistry, and Microbes." *Oceanography and Marine Biology Annual Review* 43 (2005): 1–46.

Kiel, Steffen, and James L. Goedert. "Deep-Sea Bonanzas: Early Cenozoic Whale-Fall Communities Resemble Wood-Falls Rather than Seep Communities." *Proceeding of the Royal Society* B 273 (2006): 2625–2631.

Nesbitt, Elizabeth A. "A Novel Trophic Relationship between Cassid Gastropods and Mysticete Whale Carcasses." *Lethaia* 38 (2005): 17–25.

Smith, Craig R., Adrian G. Glover, Tina Treude, Nicholas D. Higgs, and Diva J. Amon. "Whale-Fall Ecosystems: Recent Insights into Ecology, Paleoecology, and Evolution." *Annual Review of Marine Science* 7 (2015): 571–596.

Zachos, James, Mark Pagani, Lisa C. Sloan, Ellen Thomas, and Katharina A. Billups. "Trends, Rhythms, and Aberrations in Global Climate 65 Ma to Present." *Science* 292 (2001): 686–693.

Lear, Caroline H., Trevor R. Bailey, Paul N. Pearson, Helen K. Coxall, and Yair Rosenthal. "Cooling and Ice Growth across the Eocene–Oligocene Transition." *Geology* 36 (2008): 251–254.

Peredo, Carlos M., Nicholas D. Pyenson, Christopher D. Marshall, and Mark D. Uhen. "Tooth Loss Precedes the Origin of Baleen in Whales." *Current Biology* 28 (2018): 1–9.

Peredo, Carlos M., and Mark D. Uhen. "New Basal Chaeomysticete (Mammalia: Cetacea) from the Late Oligocene Pysht Formation of Washington, USA." *Papers in Palaeontology* 2 (2016): 533–554.

Pyenson, Nicholas D. *Spying on Whales: The Past, Present, and Future of Earth's Most Awesome Creatures.* New York: Penguin Books, 2019.

McCurry, Matthew R., and Nicholas D. Pyenson. "Hyper-longirostry and Kinematic Disparity in Extinct Toothed Whales." *Paleobiology* 45 (2019): 21–29.

Peredo, Carlos M., Mark D. Uhen, and Margot D. Nelson. "A New Kentriodontid (Cetacea: Odontoceti) from the Early Miocene Astoria Formation and a Revision of the Stem Delphinidan Family Kentriodontidae." *Journal of Vertebrate Paleontology* 38 (2018).

Pyenson, Nicholas D. *Spying on Whales: The Past, Present, and Future of Earth's Most Awesome Creatures.* New York: Penguin Books, 2019.

Therwissen, J. G. M. Hans. *The Walking Whales: From Land to Water in Eight Million Years.* Berkeley: University of California Press, 2014.

Berta, Annalisa, James L. Sumich, and Kit M. Kovacs. 3d ed. *Marine Mammals: Evolutionary Biology.* Amsterdam: Academic Press, 2015.

Tedford, Richard H., Larry G. Barnes, and Clayton E. Ray. "The Early Miocene Ursoid Carnivoran *Kolponomos*: Systematics and Mode of Life." Contributions in Marine Mammal Paleontology Honoring Frank C. Whitmore Jr. *Proceedings of the San Diego Society of Natural History* (1991): 11–32.

Tseng, Z. Jack, Camille Grohé, and John J. Flynn. "A Unique Feeding Strategy of the Extinct Marine Mammal *Kolponomos*: Convergence on Sabretooths and Sea Otters." *Biological Sciences Proceedings* 283, no. 1826 (2016): 1–8.

Howard, Hildegarde. "A New Avian Fossil from Kern County, California." *Condor* 71 (1969): 68–69.

Mayr, Gerald, James L. Goedert, and Olaf Vogel. "Oligocene Plotopterid Skulls from Western North America and Their Bearing on the Phylogenetic Affinities of These Penguin-Like Seabirds." *Journal of Vertebrate Paleontology* 34, no. 4 (2015).

Olson, Storrs L., and Yoshikazu Hasegawa. "Fossil Counterparts of Giant Penguins from the North Pacific." *Science* 206 (1979): 688–689.

Martin, Anthony J. *The Evolution Underground: Burrow, Bunkers and the Marvelous Subterranean World beneath Our Feet.* Cambridge, UK: Pegasus Press, 2017.

Mustoe, George E. "Enigmatic Origin of Ferruginous 'Coprolites': Evidence from the Miocene Wilkes Formation, Southwestern Washington." *Geological Society of America Bulletin* 113 (2001): 673–681.

Nelson, Derek L. "The Ravages of Teredo: The Rise and Fall of Shipworm in US History, 1860–1940." *Environmental History* 21 (2016): 100–124.

Nesbitt, Elizabeth A. "Paleoecological Analysis of Molluscan Assemblages from the Middle Eocene Cowlitz Formation, Southwestern Washington." *Journal of Paleontology* 6 (1995): 1060–1073.

Pérez, Damián E. "Phylogenetic Relationships of the Family Carditidae

(Bivalvia: Archiheterodonta)." *Journal of Systematic Paleontology* 17 (2019): 1359–1395.

Prothero, Donald R., Linda C. Ivany, and Elizabeth A. Nesbitt. *From Greenhouse to Icehouse: The Marine Eocene-Oligocene Transition.* New York: Columbia University Press, 2003.

Vermeij, Geerat J. "The Evolutionary Interaction among Species: Selection, Escalation, and Coevolution." *Annual Review of Ecology and Systematics* 25 (1994): 219–236.

Weaver, Charles E. "Paleontology of the Marine Tertiary Formations of Oregon and Washington." *University of Washington Publication in Geology* 5 (1943).

21 COILED DENIZENS OF THE SEA

Brown, Ned. *Geology of the San Juan Islands.* Chuckanut Editions. Bellingham, WA: Village Books, 2014.

Haggart, James W., Peter D. Ward, and William Orr. "Turonian (Upper Cretaceous) Lithostratigraphy and Biochronology, Southern Gulf Island, British Columbia, and Northern San Juan Island, Washington State." *Canadian Journal of Earth Sciences* 42 (2001): 2001–2020.

Landman, Neil H., Marcin Machalski, and Christopher D. Whalen. "The Concept of 'Heteromorph Ammonoids.'" *Lethaia* 54, no. 5 (2021).

22 DINOSAUR ISLAND

Peecook, Brandon R., and Christian A. Sidor. "The First Dinosaur from Washington State and a Review of Pacific Coast Dinosaurs from North America." *PLOS One* 10, no. 5 (2015).

23 THE EARLIEST ANIMAL FOSSILS

Dahla, Tais W., James N. Connelly, Da Lic, Artem Kouchinskyd, Benjamin C. Gille, Susannah Porter, Adam C. Maloof, and Martin Bizzarro. "Atmosphere–Ocean Oxygen and Productivity Dynamics during Early Animal Radiations." *Proceedings of the National Academy of Sciences* 116, no. 39 (2019): 19352–19361.

Fortey, Richard. *Trilobites: Eyewitness to Evolution.* New York: Harper Collins, 2001.

McLaughlin, Kenneth P., and Betty B. Enbysk. "Middle Cambrian Trilobites from Pend Oreille County, Washington." *Journal of Paleontology* 24 (1950): 466–471.

Kolbert, Elizabeth. *The Sixth Extinction: An Unnatural History*. New York: Henry Holt, 2014.

Okulitch, Vladimir J., and Robert G. Greggs. "Archaeocyathid Localities in Washington, British Columbia, and the Yukon Territory." *Journal of Paleontology* 32 (1958): 617–623.

Rowland, Stephen M. "Archaeocyaths: A History of Phylogenetic Interpretation." *Journal of Paleontology* 75 (2001): 1065–1078.

Wood, Rachel A. "The Rise of the First Animals." *Scientific American* 320 (2019): 33–39.

INDEX

Page references to illustrations are in *italics*.

INDEX